D0411302

30-SECOND
ASTRONOMY

30-SECOND ASTRONOMY

The 50 most mindblowing discoveries in astronomy, each explained in half a minute

Editor
François Fressin

Foreword by
Martin Rees

Contributors
Darren Baskill
Zachory K. Berta
Carolin Crawford
Andy Fabian
François Fressin
Paul Murdin

ICON

cpp

First published in the UK in 2013 by
Icon Books
Omnibus Business Centre
39–41 North Road
London N7 9DP
email: info@iconbooks.net
www.iconbooks.net

This book was conceived,
designed and produced by
Ivy Press
210 High Street, Lewes,
East Sussex BN7 2NS, UK
www.ivypress.co.uk

Creative Director **Peter Bridgewater**
Publisher **Jason Hook**
Editorial Director **Caroline Earle**
Art Director **Michael Whitehead**
Designer **Ginny Zeal**
Illustrator **Ivan Hissey**
Profiles Text **Viv Croot**
Glossaries Text **Charles Phillips**
Project Editor **Stephanie Evans**

ISBN: 978-1-84831-597-6

Printed and bound in China

Colour origination by
Ivy Press Reprographics

10 9 8 7 6 5 4 3 2 1

CONTENTS

FOREWORD
Martin Rees

The night sky is the most universal part of our environment. Throughout history, people have gazed up at the same 'vault of heaven', though each culture has interpreted it in its own way. Ever since the Babylonians, patterns have been recorded in planetary motions. The need for a precise calendar, and for navigation across the oceans, has motivated advances in timekeeping, optics and mathematics. Indeed, astronomy has always been a driver for technology. And, thanks to huge telescopes, probes and advanced computers, modern astronomers have discovered the amazing cosmic panorama described in this book.

Theorists like myself lag far behind in trying to make sense of it all. But we have made progress. We can trace cosmic history back to a mysterious beginning nearly 14 billion years ago, when everything was squeezed hotter and denser than anything that can be created in a laboratory; we understand in outline how the first atoms, stars and galaxies emerged. We realize that our Sun is a typical star among the billions in our Galaxy; and that our Galaxy is just one of many billions visible through a large telescope. Moreover, some theorists speculate that a further 'Copernican demotion' may lie ahead: physical reality is almost certainly more extensive than the domain we can observe; indeed 'our' Big Bang could be just one of many.

But recent advances haven't just extended our cosmic horizons; they have revealed richer detail. Probes to other planets of our solar system (and their moons) have beamed back images of varied and distinctive worlds. More important still, we've inferred, by detecting very slight changes in the motions and brightness of stars, that most of them are orbited by retinues of planets, just as the Earth and other familiar planets orbit the Sun. In coming years, we will be swamped by fascinating new data – perhaps even evidence of life around other stars.

Astronomy now attracts wider interest than ever before – its discoveries are part of modern culture. Moreover, the joy of discovery isn't now limited to professionals – indeed they are swamped by the

Our brightest neighbour
Planet Venus is very visible, partly because sunlight is easily reflected by the sulphurous clouds that blanket its atmosphere. Venus is the closest planet to our Earth, within reach of questing space probes, although its hot, deadly atmosphere prohibits any human exploration of its surface.

sheer quantity of data. So there is scope for 'citizen scientists', who can access and download data from surveys made with the world's best telescopes, and perhaps discover a new galaxy or a new planet. And serious amateurs, using small telescopes with the latest instrumentation, can match what professionals with much larger telescopes could do 50 years ago.

The technical details of all modern science are arcane. However, I believe that the essence of any discovery can be conveyed in accessible language. To condense a concept into 30 seconds is a bigger challenge, but one that has been triumphantly met by the authors here.

This book deserves wide readership among those fascinated by the extraordinary 'zoo' of objects in the cosmos – a cosmos governed by physical laws that allowed creatures to evolve (on Earth and perhaps on alien worlds too) with minds able to ponder its wonder and its mystery.

INTRODUCTION

François Fressin

Almost every discovery about the Universe has led us to realize how insignificant the Earth is. Compared to the rest of the Universe, our Earth represents roughly as much as a drop of water in the oceans, or a grain of sand in the deserts. In almost every field of research, astronomers have been surprised both by the extent of astrophysical structures and by their diversity.

But astronomical discoveries also tell us how strongly we are connected with the cosmos. We learn about the solar system, and we appreciate the interplay of its constituent parts with the appearance of life on Earth and its evolution. Comets brought the vast quantity of water that formed our oceans. The Moon slowed the Earth's rotation and is responsible for the tides and the seasons. Jupiter scattered asteroids that would otherwise dramatically impact on the Earth. Our connection to the stars is even stronger. The air we breathe, the iron in our blood, the carbon in our flesh; all of these came from the core of a dying star, billions of years ago.

Astronomy has this strange duality between insignificance and preciousness. A human life is so close to nothing in the immensity of space and the passage of ages. But nowhere and never has there existed someone exactly like you. If we define spirituality as an inner path enabling a person to discover the essence of his or her being, then astronomy is definitely a spiritual experience.

The topics in this book will give you an idea of this barely imaginable immensity and diversity. Unicorns, psychic powers, flying cities are fairly easy to picture. How about objects so massive that they distort space and time, a dark energy scattering the entire Universe apart? How about the fact that the breadth of your finger, held up to the sky, covers millions of galaxies, each one containing billions of stars like the Sun, or the fact that you are standing on a ball of mud that will endlessly fly through the void? Any one of those scenarios is a little tougher to imagine, but they all reflect the real world.

Giant planets
Despite their size, the very existence of Uranus and Neptune was unknown to astronomers prior to the invention of the telescope.

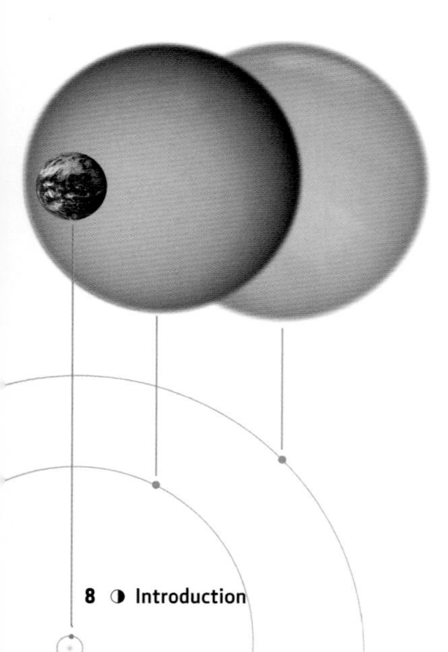

Scientists are often viewed as emotionally removed and logical people who prefer figuring out what's behind the scenes, instead of simply contemplating the beauty of the world, or who attempt to unravel a mystery rather than consider it sacred, untouchable. I believe that comprehension of the natural world does not diminish its capacity to inspire emotional wonder.

30-Second Astronomy offers 50 great astronomical discoveries, each one summed up by recognized experts in different fields of astronomical research who accepted the challenge to try to encapsulate each topic in a way that is at once succinct, accessible and reflects the consensus of the current understanding of astrophysical phenomena.

The topics are grouped in seven sections, roughly organized according to their distance from Earth and the time of their discovery. The first is **The Planets** – those other worlds next door. The second describes the other bodies of **The Solar System**, comets and asteroids, in this small part of space where everything revolves around the Sun. Section three focuses on **The Stars**, especially the dramatic ending of their lives, sometimes resulting in supernovae explosions and the formation of pulsars or black holes. In **The Milky Way** lie explanations of the objects in the night sky, and how millions of stars organize into galaxies. Section five, **The Universe**, gathers the current knowledge of what we know of the beginning of time, of the Big Bang, and of the ancestors of stars and galaxies. **Space & Time** charts the principles that rule the motions of astrophysical objects, and how much we can learn by studying the light we receive from them. The final section, **Other Worlds**, leads us back to the beginning, humans on Earth observing the sky, wondering if there are other Earths like ours – and other life. It features the recent discovery of planets orbiting stars other than the Sun. Each section profiles a herald of research in each field, summarizing the lives of exceptional scientists like Edwin Hubble or Carl Sagan.

This book serves two purposes. Its structure and approach are such that you can dip into one entry and learn what a black hole really is, or what the *Curiosity* rover is looking for on Mars. Or, read it from the start and you will have an informed overview of the state of scientific knowledge of the Universe today. Just as we don't truly know ourselves if we don't connect with others, nor do we really know the country in which we dwell unless we have travelled and lived in another country, so looking at other worlds and thinking about this Earth in the immensity of space may provide one small step towards awakening a consciousness of what inhabiting this world means.

Dramatic ending
Giant stars are far more luminous and have shorter lifespans than slower-burning dwarfs. The bigger the star, the shorter its life, which ends as a supernova explosion, leaving behind a neutron star or black hole.

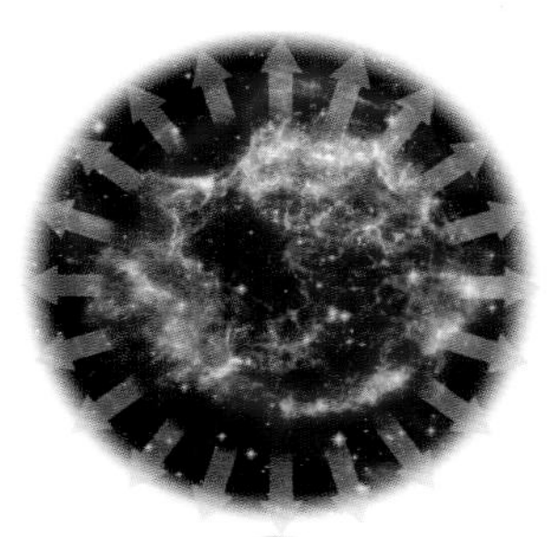

THE PLANETS

THE PLANETS
GLOSSARY

Apollo programme US National Space and Aeronautic Association (NASA) initiative to land a man on the Moon, inaugurated in 1961 and comprising 17 missions in 1967–72. *Apollo 11* made the first manned landing on the Moon's surface on 20 July 1969; *Apollo 17* was the programme's final flight in December 1972. In the course of the programme there were six lunar landings in which 12 US astronauts walked on the Moon.

atmosphere Layer of gases surrounding a planet or any body of sufficient mass, including a star; the layer's shape is maintained by gravity.

biomass Biological material that comes from living or recently living organisms.

core Central part of a planet or star.

crust The solid outer part of a planet or natural satellite.

gas giant A large planet that consists primarily of gases instead of rock. The four gas giants in our solar system are Jupiter, Saturn, Uranus and Neptune. There are other gas giants beyond the solar system, in orbit around other stars.

greenhouse effect Process through which heat radiated from a planet's surface is absorbed and then radiated outwards in all directions (including back down towards the surface) by gases in the atmosphere. As a result, the temperature on the surface and beneath the atmospheric gases is raised. Earth has a greenhouse effect, but so do other planets, such as Venus; the greenhouse effect is far stronger on Venus than on Earth.

Low Earth Orbit An orbit of the Earth at an altitude of 145–1,000 kilometres (90–620 miles). All manned space flights, apart from those of the Apollo programme, all manned space stations and most artificial satellites are in the Low Earth Orbit.

mantle A layer about 2,900 kilometres (1,800 miles) thick, between the outer part of the Earth's core and its surface (crust).

mare Areas of basaltic lava on the Moon's surface. (Basalt is a grey-to-black igneous rock.) Early astronomers wrongly identified these as areas of water and they were named *mare* (Latin for 'sea' or 'seas'). There are several such areas, including the Mare Nubium ('Sea of Clouds') and the Mare Serenitatis ('Sea of Serenity'); together, they form approximately 16 per cent of the Moon's surface. They appear as dark areas on the Moon, visible with the naked eye, and make up the patterns interpreted in some cultures as 'the Man in the Moon'.

meteor Colloquially known as a 'shooting star', the name given to a streak of light caused by rock or dust burning up as it falls through a planet's atmosphere.

meteor shower The appearance of several meteors in short succession.

meteorite A meteoroid that has landed on the surface of a moon or planet.

meteoroid A rocky body, smaller than an asteroid, in our solar system.

moon Also known as a natural satellite, an astronomical body that orbits a planet (or smaller body), known as its 'primary'. The Earth's Moon is the fifth largest natural satellite in the solar system, after Ganymede (the largest, a moon of Jupiter); Titan (second largest, a moon of Saturn); and Callisto and Io (third and fourth largest, moons of Jupiter).

niche environments Settings specialized to suit a particular species.

outgassing The release of gas that was absorbed, frozen or otherwise trapped in a surface – for example, in an ocean or area of rock on the surface of a planet.

protoplanet The 'embryos' or initial formations of planets, formed in a protoplanetary disc (cloud of dust and gas surrounding a new star). They form from the collision of smaller planetesimals. Where there are several protoplanets in orbit around a star, they collide to form one or more planets.

regolith Any loose mixture, such as soil or pieces of stone, that covers solid rock, from Greek words meaning 'blanket' and 'rock'. Found on the Earth, the Moon and on other planets, moons and asteroids.

ring system Also known as a 'planetary ring', a disc-shaped formation of dust and particles (up to several metres in size) orbiting a planet. The most celebrated ring system in our solar system is around Saturn; Neptune, Uranus and Jupiter also have planetary rings.

tectonic plates Movable pieces of a planet's crust (outer surface) and parts of the upper mantle (the layer directly beneath the crust).

tenuous Lacking in density. Used of a planet's atmosphere.

MERCURY

the 30-second astronomy

Mercury is the smallest of the eight planets, with a diameter of 4,879 kilometres (3,032 miles). The closest planet to the Sun, it is the speediest in its orbit: Mercury orbits the Sun in 88 Earth days. It rotates relative to the stars once on its axis every 59 days, turning three times on its axis for every two orbits. Because of the way that the planet rotates relative to the Sun as it orbits, its calendar is bizarre: a single day on Mercury (from sunrise to sunrise) lasts two Mercurian years or 176 Earth days. Mercury has no seasons and the largest temperature range on any planet in our solar system – from 400°C (800°F) at noon on its equator to –200°C (–300°F) near its poles at night; the temperature is especially low in the perpetually shadowed bottoms of its polar craters, where there are accumulations of ice. Mercury has a cratered, solid surface, much like the Moon. Its atmosphere is tenuous (lacking in density) and consists of atoms trapped from the Sun or outgassed from its hot surface. Mercury's craters were formed in the same way as the craters on the Moon, through bombardment by asteroids and meteors.

3-SECOND BANG
Named after the messenger of the ancient Roman gods, Mercury is a fast-moving planet of extremes – very hot by day and very cold by night.

3-MINUTE ORBIT
Mercury's orbit is the most elliptical of any planet, as well as the closest to the Sun, so it experiences a large variation in gravitational pull. This makes its orbit a test bed for the theory of gravity. Its orbit does not quite fit Isaac Newton's theory, but Albert Einstein's theory of gravity, known as General Relativity, solved the anomaly – and this was the first proof that General Relativity was better than Newton's theory.

RELATED TOPICS
See also
THE MOON
page 20

ELLIPSES & ORBITS
page 120

GRAVITY
page 124

RELATIVITY
page 126

3-SECOND BIOGRAPHY
ALBERT EINSTEIN
1879–1955
German-Swiss-American theoretical physicist

30-SECOND TEXT
Paul Murdin

With little atmosphere to act as an insulating blanket, Mercury's temperature plummets by hundreds of degrees as night falls.

VENUS

the 30-second astronomy

Venus is roughly the size of the Earth, with a diameter of 12,104 kilometres (7,521 miles). It orbits the Sun inside the Earth's orbit, once every 224 days, and rotates every 243 days – backwards. Like Earth, Venus has an atmosphere, but on Venus this is hot, dense and consists primarily of carbon dioxide, creating an intense greenhouse effect that passes on the Sun's heat to the surface and traps it below the atmosphere. As a result, the temperature on Venus averages 480°C (890°F) – hot enough to melt zinc. Seen from outside, the atmosphere supports opaque clouds that completely obscure the surface; seen from below, the sky is sulphurous yellow, as imaged by space vehicles that have landed to record the environment. Venus has been mapped by cloud-piercing radar both from Earth and from a space satellite, *Magellan* (1990–94). The surface is completely dry, and made of scaly, black volcanic rocks. Venus has more than 100 volcanoes, with solidified rivers of lava on their sides. Most terrestrial volcanoes are due to upwelling magma penetrating the surface of a planet at the edges of colliding tectonic plates – Venus has no tectonic plates and its volcanoes are fed through weak surface spots.

3-SECOND BANG

In some respects Earth's twin, the planet Venus has suffered global catastrophes that have made its surface hellish – hot, black rock beneath a sulphurous sky.

3-MINUTE ORBIT

Space vehicles sent to Venus must be strengthened to withstand the atmospheric pressure (about 90 times the pressure on Earth) and proofed against sulphuric acid rain falling from the clouds. They also have to withstand the searing heat. Landing craft that have survived the descent and landed on the rocks without falling over have operated only for an hour or so. The existence of Venusian extraterrestrials seems improbable.

RELATED TOPICS

See also
METEORS
page 48

EXTRATERRESTRIALS
page 138

3-SECOND BIOGRAPHY

CARL SAGAN
1934–96
American astronomer who identified the greenhouse effect on Venus

30-SECOND TEXT

Paul Murdin

A featureless black spot when silhouetted against the Sun during a transit, Venus has been revealed by space satellites to be a volcanic wasteland.

THE EARTH

the 30-second astronomy

3-SECOND BANG
American astronomer Carl Sagan declared of the Earth: 'That's here. That's home. That's us ... a mote of dust suspended in a sunbeam.'

3-MINUTE ORBIT
Since the first satellites took photographs of the Earth, it has often been called 'the blue planet' on account of the dominance in these images of blue oceans. However, water represents only 0.02 per cent of the Earth's mass; the oceans are like a thin sheet of blue paper covering a brown ball. And only 0.001 per cent of Earth's water is accessible fresh water, on which millions of species – *Homo sapiens* included – rely.

The Earth is a dense ball of iron and rock, the largest solid body in our solar system. It formed 4.5 billion years ago, from a mass of dust and gas left over from the formation of the Sun. Although the assembly of the Earth was practically completed within 10 million years, it is accurate to say that its formation is still not finished, because the Earth remains geologically active. The Earth's crust is divided into 15 segments called tectonic plates between 5 and 50 kilometres (3 and 30 miles) thick, all slowly floating over a silicate mantle. Beneath the mantle lies an iron-nickel core. The Earth's surface, a thin layer of a small, rocky body orbiting a typical star in the suburbs of an average galaxy, is to date the only place we know to harbour life. Life appeared on Earth within its first billion years and developed to encompass millions of species. Vegetation is the dominant life form on Earth, at least in terms of biomass and environmental impact. It has altered the atmospheric composition of the Earth and could potentially be detected from distant space by its characteristic reflectivity in infrared.

RELATED TOPICS
See also
SUPER-EARTHS & OCEAN PLANETS
page 146

TOWARDS ANOTHER EARTH
page 148

3-SECOND BIOGRAPHY
CARL SAGAN
1934–96
American astronomer, astrophysicist and writer

30-SECOND TEXT
François Fressin

Planet Earth is a ball of mud and metal, covered by a thin layer of water.

THE MOON

the 30-second astronomy

3-SECOND BANG
The Moon is the natural satellite of the Earth, and the most distant place to which any man or woman has travelled.

3-MINUTE ORBIT
The Moon has no atmosphere, and its surface is covered by meteor impacts. Mostly on the side visible from the Earth, the lunar maria ('seas') are, in fact, basaltic plains formed by ancient volcanic eruptions. Very small particles of crust called regolith cover the Moon's surface and make it as reflective as charcoal. The Moon's distance from the Earth varies in the course of its orbit: the average is 384,400 kilometres (238,900 miles).

Despite being a common sight for us all, the Moon is one of the most peculiar objects in our solar system. It is the fifth largest satellite in the solar system, and the largest one relative to its host planet. The Moon likely formed as a result of the giant impact of a Mars-size body hitting the newly formed Earth. The scattered remains of both the Earth and the impactor formed the body of the Moon. After it started orbiting the Earth, the Moon slowly drifted into a more distant orbit; its rotation synchronized, which means it rotates around its axis in the same time it takes to orbit the Earth and always shows the same face to the Earth. The Moon's most noticeable influence on the Earth is the tidal force that causes the Earth to dilate in the direction of the Moon, due to the stronger gravitational force affecting the part of the Earth that is closest to the Moon, and results in elevated ocean tides. Only 21 humans have gone further than the Low Earth Orbit, all in 1969–72 during the *Apollo* lunar programme, offering a glimpse of what a spatial civilization could be.

RELATED TOPICS
See also
THE EARTH
page 18

3-SECOND BIOGRAPHIES
NEIL ARMSTRONG
1930–2012
Former NASA astronaut, the first man to walk on the Moon

EDWIN EUGENE 'BUZZ' ALDRIN
1930–
Former NASA astronaut, the second man to walk on the Moon

30-SECOND TEXT
François Fressin

Buzz Aldrin, the lunar module pilot on* Apollo *11, and the second man to set foot on the Moon, described its landscape as a 'magnificent desolation'.

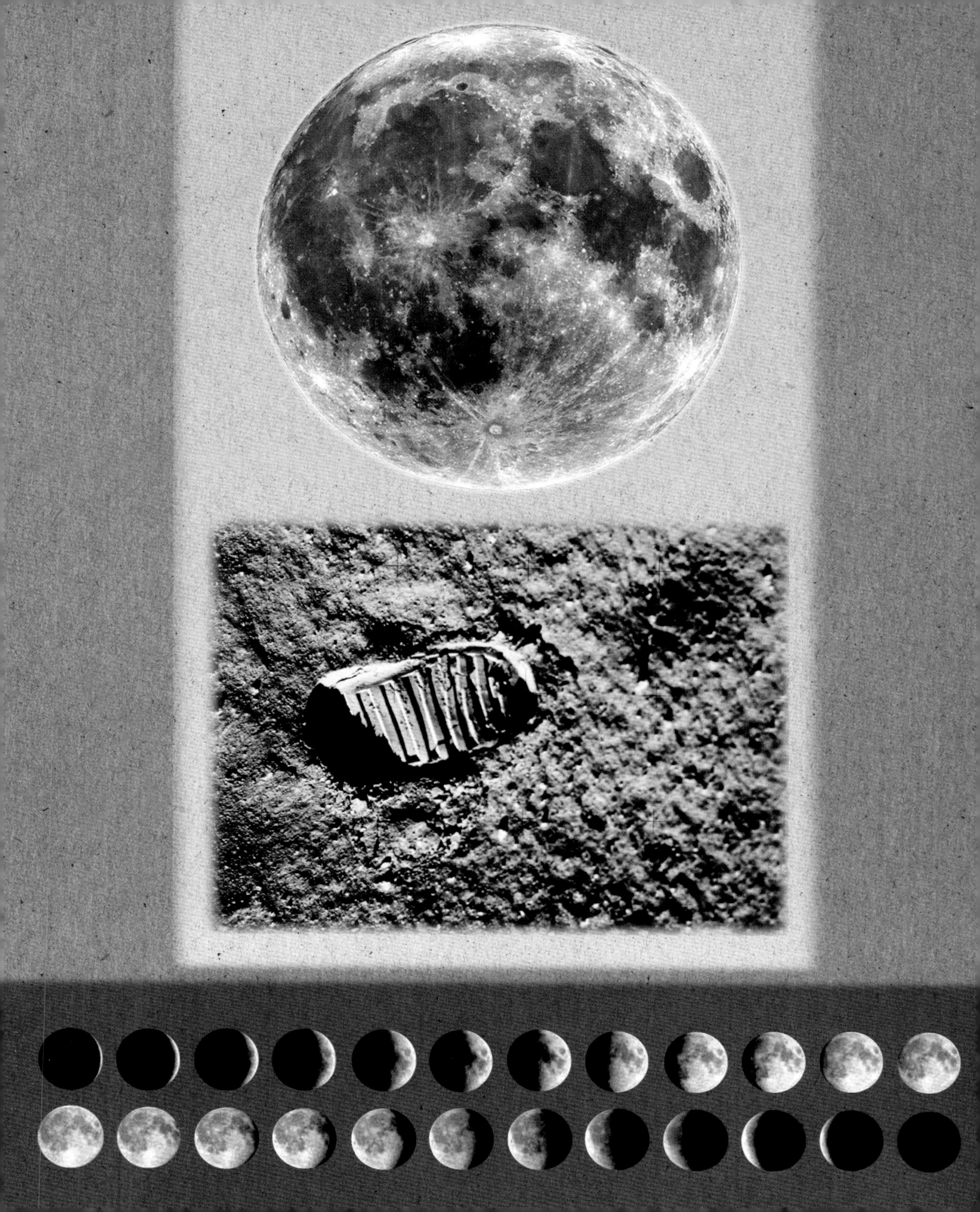

MARS

the 30-second astronomy

The next planet beyond Earth moving outwards from the Sun, Mars has a year 687 Earth days long, and it rotates in just over 24 hours. With a diameter of 6,792 kilometres (4,221 miles), Mars is smaller than Earth, but boasts both the largest mountain in the solar system – the 22,000-metre (70,000-ft) high volcano Olympus Mons – and a canyon system, the Valles Marineris, with dimensions that in parts are ten times those of the Grand Canyon in Arizona. Mars has polar caps of layered water ice and dry ice (solid carbon dioxide) that wax and wane during its seasons. Frost forms on parts of its surface during cold nights, disappearing in the morning sunlight. Although its atmosphere is thin, surface winds blow up dust storms that can envelop the entire planet. In places, thin trickles of water seep down cliff faces from ice melting below the surface. In the past, water was more abundant on Mars. There were lakes in some meteor craters, and there remain flood plains of rounded boulders, scoured by a surge of water released when an ice dam collapsed. Conditions on Mars are extreme, but this planet offers hope of finding extraterrestrial life in niche environments.

3-SECOND BANG
Mars is the planet most like Earth in our solar system, with ice caps, desert plains, mountain ranges, volcanoes and wide, deep canyons.

3-MINUTE ORBIT
Mars has a weak magnetic field, but – as revealed by residual magnetism in its old rocks – its magnetic field was stronger in the past. A planet's magnetic field is caused by the circulation of its liquid iron core. A global catastrophe dried out Mars. This was triggered by the loss of its magnetic field when its small iron core froze, which allowed the solar wind to penetrate to Mars' atmosphere and erode it.

RELATED TOPICS
See also
THE SOLAR WIND
page 38

EXTRATERRESTRIALS
page 138

3-SECOND BIOGRAPHY
PERCIVAL LOWELL
1855–1916
American astronomer, founder of Lowell Observatory (Flagstaff, Arizona) to study Mars

30-SECOND TEXT
Paul Murdin

The* Sojourner *rover (11 kg/25 lb), which explored the surface of Mars in 1997, is dwarfed by the* Curiosity *rover (1 ton/ 2000 lb), which landed on the planet in 2012.

JUPITER

the 30-second astronomy

Cold Jupiter lies five times further from the Sun than does the Earth, and takes 11.86 Earth years to complete a single orbit. It contains more than twice the combined mass of all the other planetary bodies of the solar system. Despite having a volume more than 1,300 times that of Earth, Jupiter rotates once in less than 10 hours, leaving it slightly squashed at the poles. Jupiter is not a solid body, but is composed of the lightest elements in the Universe, predominantly hydrogen and helium. The 'surface' of Jupiter we see is only the tops of the clouds in the upper regions of its gaseous atmosphere; deeper down, the gas is progressively compressed by the weight of overlying layers to become hotter and denser, until a liquid hydrogen layer surrounds a rocky core ten times the mass of the Earth. Atmospheric motions powered by solar energy and internal heat stir up complex weather patterns in the clouds, which are wrapped around the planet by its fast rotation to form colourful bands parallel to the equator. Many storms come and go, but are dwarfed by the Great Red Spot, a hurricane large enough to swallow two Earths.

3-SECOND BANG
The largest planet in our solar system, Jupiter is a world unlike our own; with its deep gaseous atmosphere, it is the archetypal gas giant.

3-MINUTE ORBIT
Jupiter has an array of more than 60 moons. The four largest – Ganymede, Callisto, Io and Europa – were discovered in 1610 by Galileo Galilei, and his observations of their movement around the giant planet helped to convince him that the Sun instead of the Earth lies at the centre of the solar system. The four largest moons are similar to our Moon in size, but all the others are far smaller and irregular in shape.

RELATED TOPICS
See also
GALILEO
page 26

SATURN
page 28

URANUS & NEPTUNE
page 30

3-SECOND BIOGRAPHY
GALILEO GALILEI
1564–1642
Italian astronomer

30-SECOND TEXT
Carolin Crawford

With a diameter of 142,700 kilometres (88,700 miles) across its equator, Jupiter dwarfs planet Earth.

15 February 1564
Born in Pisa

1581
Studied medicine at the University of Pisa

1586
Invented the hydrostatic balance

c. 1592
Invented the thermoscope

1592–1610
Taught mathematics, mechanics and astronomy at the University of Padua

1610
Published *Sidereus Nuncius* (*The Starry Messenger*), a brief treatise describing his telescopic observations

1612
Observed Neptune, but did not recognize that it was a planet

1616
First account of the motion of the tides in *Discorso del Flusso e Reflusso del Mare* (*Dialogue on the Ebb and Flow of the Sea*), the basis for the later *Dialogue*

1616
Observed Saturn's rings

1616
Defended heliocentrism at the Roman Inquisition

1617
Observed the double star Mirzar in Ursa Major

1623
Published *Il Saggiatore* (*The Assayer*)

1632
Published *Dialogo dei due Massimi Sistemi del Mondo* (*Dialogue on the Two Chief World Systems*), a defence of heliocentric theories

1633
Roman Inquisition finds him guilty of heresy; he is put under house arrest

1634–38
Wrote *Discorsi e Dimostrazioni Matematiche, intorno a due Nuove Scienze* (*Discourses and Mathematical Demonstrations Relating to Two New Sciences*), a summary of his work on strength of materials and the geometry of motion

1638
Became blind

8 January 1642
Died in Florence

1718
Ban on the printing of his books is lifted

1835
His works were removed from the Catholic Church's Index of Forbidden Books

GALILEO

Mathematician, astronomer, physicist, artist, musician, teacher, physician, inventor, writer: Galileo Galilei was the late Renaissance man's Renaissance man. The son of a musician, and no mean musician himself, he started out as a medical student in Pisa, but was soon seduced by mathematics and physics, then got sidetracked by art and design. Throughout his life, Galileo was oppressed by family financial problems and was constantly looking to invent money-spinning products – including the thermoscope, a forerunner of the thermometer and a military compass.

Although he was a major player in the seventeenth-century scientific revolution (Albert Einstein called him the father of modern science, and he is well known for his work in physics, on falling bodies) we know him best as an astronomer and mapper of the Moon. He made improvements on the telescope developed by German-Dutch lensmaker Hans Lippershey (1570–1619) and observed, identified and noted the phases of Venus and the four largest moons of Jupiter, as well as sunspots, and also discovered that the Milky Way was composed of billions of stars. These first-ever observations by telescope were recorded in his treatise *The Starry Messenger* (1610).

Galileo was a committed supporter of the heliocentric theories of Copernicus, which proposed that the Earth, the Moon and the planets revolved around the Sun; Galileo used his observations to support this theory after the Roman Inquisition had concluded in 1616 that the heliocentric structure was impossible. Despite being warned against taking this action, Galileo – who had a short fuse, a caustic wit and a healthy disregard for authority – then published *Dialogue on the Two Chief World Systems* in 1632. He had already challenged authority and promoted experimentation in his 1623 work *The Assayer* (now regarded as his scientific manifesto), which had enjoyed success, but the *Dialogue* was seen as an insult to the increasingly paranoid Pope Urban VIII, and the wrath of the Catholic Church fell on Galileo's head. He was tried by the Inquisition, and found 'vehemently suspect of heresy'; under the threat of torture, he reluctantly recanted, but was put under house arrest for the rest of his life and his work added to the Index of Forbidden Books. The world had to wait until the early nineteenth century to read Galileo's works and find him vindicated.

SATURN

the 30-second astronomy

A gas giant, and the second largest planet in the solar system, Saturn has a volume equivalent to that of more than 700 Earths, but its mass is only 95 times greater. It has the lowest density of all the planets – less dense than water on Earth. Its deep atmosphere is composed of hydrogen and helium wrapped around a small rocky core; the atmosphere is squashed outwards by fast rotation to make it 10 per cent wider at the equator than at the poles. Differences in the observed spin rate of the planet's features suggest that a day on Saturn is 25 minutes longer at the poles than at the equator. Saturn has more than 60 moons – ranging from tiny moonlets less than a kilometre across to the gigantic Titan, which has a diameter of 5,150 kilometres (3,200 miles) and is larger than the planet Mercury. Titan is comparable to Earth: it has a layered atmosphere, and is the only other known object in our solar system with stable surface liquid. Many of Saturn's smaller moons, such as Phoebe, have orbital characteristics that suggest that they were originally asteroids that have been captured by Saturn's gravity.

3-SECOND BANG
Saturn – the furthest of the planets known to humankind since antiquity – is best known for its spectacular ring system and varied retinue of moons.

3-MINUTE ORBIT
Saturn's rings are composed of chunks of ice and rock, the debris of a small moon shattered by gravitational forces 100 million years ago. Extending about 6,400–120,700 kilometres (4,000–75,000 miles) above Saturn's equator, the rings are only 100 metres (325 ft) thick. During Saturn's 29½ year journey around the Sun, the rings' inclination changes as viewed from Earth, altering the apparent shape from wide open when face-on to almost invisible edge-on.

RELATED TOPICS
See also
JUPITER
page 24

ASTEROIDS
page 42

3-SECOND BIOGRAPHIES
GIOVANNI CASSINI
1625–1712
French-Italian astronomer who discovered four of Saturn's moons

CHRISTIAAN HUYGENS
1629–95
Dutch astronomer who discovered Titan

30-SECOND TEXT
Carolin Crawford

Saturn's rings are not solid, but are trillions of particles of rock and ice in orbit around the planet. They range from the size of sand grains to that of small boulders.

URANUS & NEPTUNE

the 30-second astronomy

The most remote planets in our solar system, Uranus and Neptune orbit the Sun at distances respectively 19 and 30 times further than that of the Earth. Consequently, they are both frigid worlds, with mean temperatures in their atmospheric clouds of about –200°C (–300°F) and take, respectively, 84 and 165 Earth years to orbit the Sun once. They have been visited by one spacecraft, *Voyager 2*, which flew past Uranus in 1986 and Neptune in 1989. Both planets have a faint ring system, and are accompanied by a coterie of moons. Their deep, relatively featureless atmospheres are dominated by hydrogen and helium and surround a large core of rock and ice. Extra hydrocarbons, such as ammonia and methane, change the colour of the sunlight reflected by the cloud tops to give the planets their distinctive green and blue hues. Internally generated heat stirs up some of the fastest winds in the whole solar system on Neptune, reaching speeds of up to 2,000 km/h (1,250 mph). Uranus is tipped over to rotate completely on its side, an unusual orientation probably caused by a collision with another protoplanet soon after its formation.

3-SECOND BANG
Uranus and Neptune are the two outermost giant gas planets of our solar system, each with a diameter about four times that of Earth.

3-MINUTE ORBIT
Uranus and Neptune are the only planets in our solar system to have been discovered in modern times, and by telescope. Uranus was discovered serendipitously in 1781 by William Herschel; from observed anomalies in its motion around the Sun astronomers deduced that it was experiencing the gravitational pull of a more distant planet. John Couch Adams and Urbain Le Verrier independently predicted its location, leading to the discovery of Neptune by Johann Galle in 1846.

RELATED TOPICS
See also
JUPITER
page 24

SATURN
page 28

3-SECOND BIOGRAPHIES

WILLIAM HERSCHEL
1738–1822
German-born British astronomer, discoverer of Uranus

URBAIN LE VERRIER
1811–77
French mathematician and astronomer

JOHANN GOTTFRIED GALLE
1812–1910
German astronomer, discoverer of Neptune

JOHN COUCH ADAMS
1819–92
British mathematician and astronomer

30-SECOND TEXT
Carolin Crawford

Uranus and Neptune are the outermost giant worlds of our solar system.

THE SOLAR SYSTEM

THE SOLAR SYSTEM
GLOSSARY

astronomical unit One astronomical unit (AU) is the mean distance between the Earth and the Sun: roughly 150 million kilometres (93 million miles). Towards the edge of the solar system, the Kuiper Belt (containing dwarf planets) is 30–55 AU from the Sun, while the far more distant Oort Cloud (containing icy objects) is 5,000–100,000 AU from the Sun.

coma Very thin cloud of gas and dust surrounding the heart or nucleus of a comet. The heart is a ball of ice and rock particles, described by American astronomer Fred Whipple as an icy conglomerate, or 'dirty snowball'. As the comet nears the inner solar system and is more strongly heated by the Sun, some of the ice and dust is vaporized, creating the coma.

convective zone In the Sun, an area between the radiation zone (nearer the core) and the solar photosphere, through which energy passes by convection. Hotter material rises from the bottom, carrying energy, then sinks again after cooling; the cooling material heats up again as it sinks and then rises once more in a rolling process.

Halley's Comet Officially known as 1P/Halley, a short-period comet named after English astronomer Edmond Halley, who correctly calculated in 1705 that the comets seen in 1531, 1607 and 1682 were one returning comet and that this comet would return in 1758. Halley is the brightest short-period comet visible to the naked eye and is visible every 75–76 years. It has been known since at least 240 BC, was seen during the Norman Conquest of England in 1066, and was represented in the Bayeux Tapestry that recorded the conquest. Last seen in 1986, it will appear again in 2061.

Kuiper Belt Doughnut-shaped region in the outer solar system, billions of kilometres from our Sun, containing small bodies and dwarf planets, including Pluto. Because their orbit lies beyond that of Neptune, they are often called 'trans-Neptunian objects'.

nuclear fusion The combination (fusion) of two atomic nuclei to form a heavier nucleus, accompanied by the release of energy. Nuclear fusion powers the Sun and other active stars.

Oort Cloud Spherical cloud in the outer solar system, far beyond the Kuiper Belt, that could contain up to two trillion frozen bodies. The further reaches of the Oort Cloud mark the limit of the Sun's gravitational attraction – and so are at the boundary of our solar system. Astronomers believe that most comets originate in the Oort Cloud.

orbital period Time taken for an object to make a complete orbit around another. The Earth's orbital period around the Sun is one year, or 365.256363 days.

Perseids The meteor shower occurring annually from 23 July to 20 August, so called because the area from which the meteors appear to fall lies in the constellation Perseus. The dust and debris come from the comet Swift-Tuttle. The Perseids are mostly visible in the northern hemisphere.

protoplanetary disc Rotating disc of gas and dust surrounding a newly formed star in a developing solar system. Planets form from the gas and grains of dust.

short-period comet Comet with an orbital period around the Sun of less than 200 years.

solar corona The outer atmosphere of the Sun, not normally visible because it is one million times less bright than the visible solar photosphere. The corona can be seen during a total solar eclipse, when the brightness of the solar disc is blocked by the Moon, or using a coronagraph instrument, which blocks the light coming from the solar disc in order to enable study of the solar atmosphere.

solar photosphere The visible outer layer of the Sun, only about 100 kilometres (60 miles) thick. Sunspots, faculae (bright areas) and granules (cellular features) are visible on the photosphere.

star A vast ball of gas of great mass (held together by gravity) that generates heat and light through nuclear fusion reactions at its core.

sunspots Dark spots on the solar photosphere that result when magnetic activity limits convection, creating areas where the very high temperature is partly reduced.

THE SUN

the 30-second astronomy

Very high temperatures and pressures deep in the core of the Sun squeeze hydrogen into helium, converting a fraction of the atoms' mass into pure energy through nuclear fusion. This energy radiates outwards and bubbles, like water in a boiling pot, up through the Sun's convective zone, riding on buoyant plumes of ionized gas (plasma). Finally, having travelled 700,000 kilometres (435,000 miles) from the centre of the Sun (100 times further than the distance from the Earth's core to its surface), the energy escapes as bright white light from the solar photosphere (the visible outer layer of the Sun), then radiates into the darkness of space. Just one-billionth of this energy actually lands on the Earth, where it drives our weather and keeps us warm. Although it is the coolest layer of the Sun, the solar photosphere is so hot – at 5,500°C (10,000°F) – that it would vaporize any solid material. Strong magnetic fields, generated by swirling eddies of plasma deep in the convective zone, pierce the photosphere and leave behind dark sunspots that mottle the Sun's surface. Sunspots are most prevalent close to times of maximum solar magnetic activity, which occur once every 11 years.

3-SECOND BANG

Our star, the Sun, is a 100 trillion terawatt nuclear furnace, the source of energy for almost every living organism on Earth.

3-MINUTE ORBIT

The Sun is a star, just like the 100,000,000,000 other stars in our Galaxy. Because the Sun is so close, we can study it better than any other star. In 'helioseismology', scientists use slow oscillations seen on the Sun's surface to measure its internal structure and composition. Using underground detectors, they can even observe weakly interacting fundamental particles (neutrinos) that are by-products of the nuclear reactions taking place in the Sun's core.

RELATED TOPICS

See also
THE EARTH
page 18

THE SOLAR WIND
page 38

COLOUR & BRIGHTNESS OF STARS
page 54

THE LIGHT SPECTRUM
page 122

3-SECOND BIOGRAPHIES

JOSEPH VON FRAUNHOFER
1787–1826
German optician

JOSEPH NORMAN LOCKYER
1836–1920
English scientist and astronomer

30-SECOND TEXT

Zachory K. Berta

The Sun and the Earth are shown with their relative sizes (but not distances) to scale. A million Earths could fit inside the Sun.

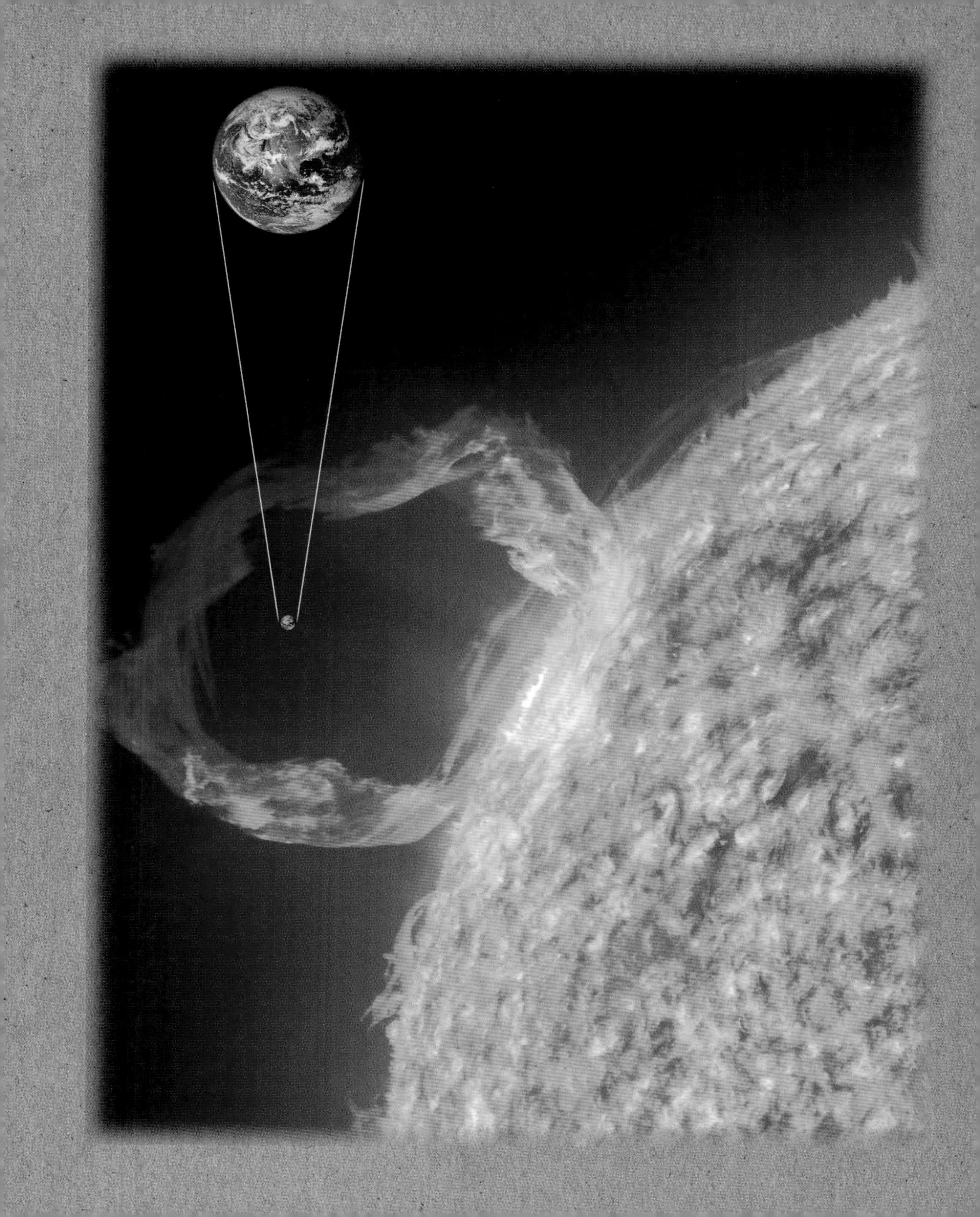

THE SOLAR WIND

the 30-second astronomy

3-SECOND BANG
In addition to emitting light, the Sun also spews out the solar wind – a supersonic blast of charged particles constantly buffeting the Earth's magnetic field.

3-SECOND ORBIT
In 1859, solar astronomer Richard Carrington witnessed a bright flash of light on the Sun's surface. This massive solar flare launched a solar wind gust that smashed into Earth's magnetic field one day later, creating colourful night-time aurorae bright enough to read the newspaper by and shocking telegraph operators' fingers with electrical sparks. Another geomagnetic storm as intense as this 'Carrington event' could have crippling effects on our modern telecommunications and power grids.

Above the solar photosphere hovers a tenuous billow of even hotter plasma, known as the solar corona. What heats this corona is still an active area of research, but the process may involve magnetic or acoustic waves crashing above the Sun's surface. This corona exuberantly launches billions of tons per hour of energetic particles (electrons, protons and heavier ions) out into the vacuum of space, blowing a 'solar wind' outwards at millions of kilometres per hour. At these speeds, when a solar flare causes an outburst in the solar wind above usual levels, it can travel from the Sun to the Earth in a few days. Fortunately, our intrinsic magnetic field shields us from this dangerous wind, safely deflecting it and preventing it from destroying our satellites or obliterating our biosphere. Deflected charged particles from the solar wind spiral along the Earth's magnetic field towards the poles, where they interact with our atmosphere to create the colourfully glowing *aurora borealis* ('Northern Lights') or *aurora australis* ('Southern Lights'). Aurorae grow brighter and extend further towards the equator when the Sun is most active and the solar wind most intense. The solar wind also sculpts the straight plasma tail of many comets.

RELATED TOPICS
See also
THE SUN
page 36

COMETS
page 46

3-SECOND BIOGRAPHIES
RICHARD CARRINGTON
1826–75
English astronomer

KRISTIAN BIRKELAND
1867–1917
Norwegian scientist who identified the cause of the Northern Lights

30-SECOND TEXT
Zachory K. Berta

Particles from the solar wind interact with Earth's magnetic field atmosphere, creating beautiful aurorae that are more commonly seen closer to Earth's north or south poles.

ERIS, PLUTO & DWARF PLANETS

the 30-second astronomy

3-SECOND BANG
Dwarf planets could be called the adolescent offspring of the main planets – miniature versions of their seniors, without dominance over the space they inhabit.

3-MINUTE ORBIT
Planets were shaped by gravity as their material settled down over time – the interior of each planet supports its own weight. This is true both of the main planets and dwarf planets that are larger than about 560 kilometres (350 miles) in diameter, depending on composition and rotation. The main planets have additionally cleared their neighbourhood either by absorbing or ejecting anything that has strayed too close. Dwarf planets, by contrast, do not dominate their orbital zone.

Beyond the main planets lies a zone of smaller ones called trans-Neptunian objects. They orbit the Sun in an outer region of the solar system called the Kuiper Belt. They are a motley collection of planetary scraps ejected from the inner solar system by the combined and repeated tug of Jupiter and Saturn early in the solar system's development. Eris and Pluto, two of these trans-Neptunian objects, are respectively the ninth and tenth most massive planets in the solar system: Eris is 2,325 kilometres (1,445 miles) in diameter and Pluto almost identical: 2,320 kilometres (1,440 miles). Pluto was considered the ninth planet when it was identified in 1930. However, the discovery of Eris in 2005 provoked astronomers to rethink the definition of a planet, and in linked decisions in 2006 and 2008 Eris and Pluto were reclassified as 'dwarf planets'. Ceres (the largest asteroid in the main belt of asteroids between Mars and Jupiter, discovered in 1801) was also reclassified as a dwarf planet. Further trans-Neptunian objects were discovered and classified as dwarf planets between 2002 and 2007: Haumea, Makemake, Orcus, Quaoar and Sedna, all named after mythological creatures from the creation myths of less well-known civilizations.

RELATED TOPICS
See also
ASTEROIDS
page 42

3-SECOND BIOGRAPHIES
GERARD KUIPER
1905–73
Dutch-American astronomer and planetary scientist

CLYDE TOMBAUGH
1906–97
American astronomer who discovered Pluto

30-SECOND TEXT
Paul Murdin

Pluto and its four small satellites go around the Sun scarcely further out than planet Neptune, and in an inclined, highly elliptical orbit that suggests its erratic journey from the inner solar system.

ASTEROIDS

the 30-second astronomy

Our solar system hosts only eight major planets, but is swarming with a multitude of smaller rocky asteroids; hundreds of thousands have been observed so far. These airless chunks of rock and metal come in all sizes, from misshapen specks of dust to the dwarf planet Ceres, almost 1,000 kilometres (600 miles) in diameter. Asteroids exist throughout the solar system, from inside Mercury's orbit to outside Neptune's, but they only survive for more than a short period where the gravitational forces exerted by the more massive planets do not sweep them away or suck them into a collision course. Asteroids are more common in the main asteroid belt, just beyond the orbit of Mars. The spaces between asteroids are so huge, even in this asteroid belt, that asteroids collide with each other quite infrequently. Except for these rare collisions, many asteroids have remained unchanged since they condensed out of the primordial protoplanetary disc 4.5 billion years ago, during the birth of the solar system. They preserve a record of ancient conditions that we can study with telescopes, by visiting them with spacecraft, or when chunks of asteroids fall to Earth as meteorites.

3-SECOND BANG
Rocky asteroids litter our solar system and hold clues to its beginnings and evolution.

3-MINUTE ORBIT
Asteroids whose orbits bring them close to our planet are known as Near Earth Objects (NEOs). The probability of a large, life-threatening asteroid colliding with the Earth is small, but astronomers carefully track NEOs because the consequences of a collision would be so catastrophic. The Minor Planet Center of the International Astronomical Union compiles data and calculates orbits for all known asteroids, so that it can predict any potentially hazardous collisions before they occur.

RELATED TOPICS
See also
ERIS, PLUTO & DWARF PLANETS
page 40

COMETS
page 46

METEORS
page 48

3-SECOND BIOGRAPHIES
GIUSEPPE PIAZZI
1746–1826
Italian astronomer who discovered the asteroid Ceres

DANIEL KIRKWOOD
1814–95
American astronomer who identified the 'Kirkwood gaps' in the asteroid belt

KIYOTSUGU HIRAYAMA
1874–1943
Japanese astronomer who first found groups of asteroids sharing almost identical orbits

30-SECOND TEXT
Zachory K. Berta

Asteroids are tiny in size compared to the vast distances that lie between them.

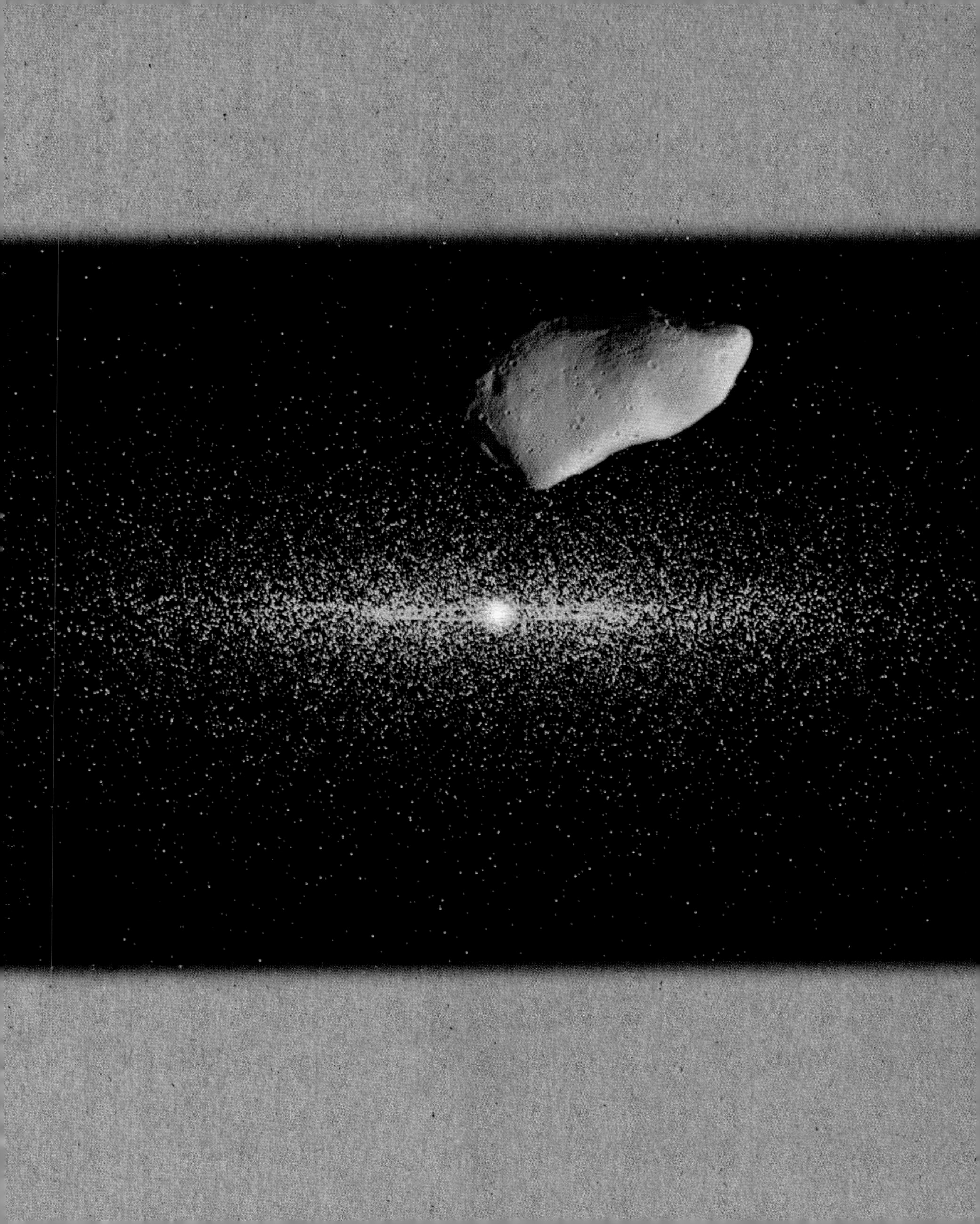

19 February 1473
Born in Torun (Thorn), now in Poland

1491–95
Studied mathematics, astronomy and natural science at the University of Cracow

1495
Elected canon, but his installation postponed

1496–1501
Studied canon law at the University of Bologna; became assistant to Italian astronomer Domenico Maria de Novara

1497
Formally appointed canon

1501–03
Studied medicine at the University of Padua

1503
Received a doctorate in law from the University of Ferrara

1503–10
Secretary and physician to his uncle, the Prince-Bishop of Warmia

c. 1514
Wrote the *Commentariolus*, an initial outline of his heliocentric theory

1512–15
Made observations of Mars, Saturn and the Sun

1532
De Revolutionibus Orbium Coelestium (*On the Revolutions of the Heavenly Spheres*) is almost finished, but Copernicus is reluctant to publish for fear of scornful reaction

1533
Johann Albrecht Widmanstetter lectured on the Copernican theory and was heard by Pope Clement VII; Copernicus was urged to publish but still reluctant

1539
Mathematician Georg Joachim Rheticus visits Copernicus and becomes his pupil and secretary

1540
Rheticus writes and publishes *Narratio Prima*, a description of the Copernican theory, which tested reactions to the idea

1542
Rheticus takes the manuscript of *De Revolutionibus* to Nuremberg

1543
De Revolutionibus is published

24 May 1543
Dies at Frombork

1566
Second edition of *De Revolutionibus* is published

COPERNICUS

Nicolaus Copernicus – the man

who turned the world upside down, or at least turned the solar system inside out – was an unlikely revolutionary. His paradigm-shifting heliocentric theory stated that the Sun, not the Earth, was at the centre of the Universe. This challenged the orthodoxy of the Ptolemaic system, named after the Egyptian astronomer Ptolemy of Alexandria (c. AD 100–c. AD 170), which stated that the Sun, Moon and planets revolved around the Earth. Copernicus's theory was painstakingly put together over a long period (no one knows when he started, but it was probably in about 1510) and was the result of early pickings at mathematical threads dangling from the edges of the Ptolemaic system. Why did the planets not follow uniform concentric orbits? And why was the theory of the equant (a point Ptolemy had invented to make everything work) so unsatisfactory? Once he had started picking, Copernicus carried on until he had unravelled the whole thing, and he came to the conclusion that the maths only worked if the Sun, not the Earth, were at the centre of everything.

Multitalented Renaissance polymath – scholar, linguist, translator, mathematician, astronomer, physician, artist, economist, diplomat and cleric – Copernicus was above all a dutiful family man. His education and career were paid for and directed by his maternal uncle Lucas Watzenrode, the powerful Bishop of Warmia, who had determined that Copernicus should achieve high office in the Church. This the young man did: after studies at the University of Cracow, he was elected canon at Frombork (Frauenberg), in northern Poland, where he stayed for the rest of his life, taking several leaves of absence for further study at Padua, Bologna and Ferrara. His administrative duties and the medical care of his benefactor took up much of his time, and he had to fit astronomical observations in whenever he could.

Copernicus wrote three astronomical works: *Commentariolus* (The Little Commentary), a 40-page précis of what was to become his heliocentric hypothesis, written before 1514 and circulated among peers and friends; the 'Letter against Werner' (1524), a critical drubbing of the work of mathematician Johann Werner; and his great *De Revolutionibus Orbium Coelestium* (*On the Revolutions of the Heavenly Spheres*), published in the year of his death, 1543. According to legend, Copernicus was on his deathbed when an advance copy of the printed book arrived: it was placed in his hands just before he died.

COMETS

the 30-second astronomy

Far from the Sun, a ball of ice and rock, only a few kilometres across, drifts slowly through the frozen dark. After spending most of its life as an inert, dirty snowball, it gradually accelerates inwards. The increasing warmth of the Sun heats its outermost surface, vaporizing ice to form a diffuse, gaseous 'coma' tens of thousands of kilometres in size. In this spewing state, it is an active comet and will release increasing amounts of material as it plummets further into the hot inner solar system. Two tails stretch millions of kilometres out from the coma: a curved, yellowish tail, from sunlit dust blasted off the nucleus and now drifting slowly behind; and a straight, bluish tail pointed away from the Sun, from plasma trapped in the magnetized solar wind. Comets can occur from any ice-rich bodies, be they once-in-human-history visitors from the distant Oort Cloud or Kuiper Belt objects on shorter eccentric orbits, such as Halley's Comet that returns every 75–76 years. The latter may eventually become inactive asteroids, as outbursts over successive visits past the Sun strip them of their comet-fuelling volatiles.

3-SECOND BANG
The beautiful comets that glitter across the night sky are not static, stable or constant objects but rather transient events, evolving processes – celestial happenings.

3-MINUTE ORBIT
Giant collisions during Earth's formation may have blasted the oceans and atmosphere off our young planet. Outgassing of volatile molecules trapped in Earth's mantle could provide much of the missing water and gas, but astronomers also believe that bombardment by water-rich comets may have contributed to Earth's hydrosphere. Without this substantial surface water or a protective atmosphere, it would have been difficult for life to arise on our planet.

RELATED TOPICS
See also
ERIS, PLUTO & DWARF PLANETS
page 40

ASTEROIDS
page 42

3-SECOND BIOGRAPHIES
EDMOND HALLEY
1656–1742
English astronomer who first computed the orbit of Halley's Comet

JAN OORT
1900–92
Dutch astronomer who gave his name to the Oort Cloud of comets

FRED WHIPPLE
1906–2004
American astronomer who first described the nuclei of comets as icy conglomerates, or 'dirty snowballs'

30-SECOND TEXT
Zachory K. Berta

The nucleus of a comet is tiny, just a millionth the comet's total size.

METEORS

the 30-second astronomy

Be it a fleck off an asteroid, a pebble from a comet's tail, or even man-made trash, a 'meteoroid' is any small body out in space that will eventually collide with our Earth. Once it enters the Earth's atmosphere, typically with speeds of 10–70 km/s (7–44 miles/s), it becomes a 'meteor'. Air friction slows its rapid descent and heats it to the point of incandescence, creating a burning streak across the sky. A chunk of rock or metal, a nugget of the original meteoroid, may survive all the way to the ground to become a 'meteorite'. Meteors can be seen on clear nights, but occur most often during annual meteor showers (for example, the Perseids in early August), when the Earth's orbit brings it through the meteoroid-rich debris clouds left by long-gone comets. Those meteors that we see flash across the sky are typically from meteoroids about 1 cm (½ in) in diameter. 'Micrometeors' (10–100 microns in diameter) are vastly more common, but fall unnoticed because they are so small. The largest meteors are extraordinarily rare. An example is the 'Cretaceous-Paleogene impactor', 10 kilometres (6 miles) in diameter, whose collision with the Earth 65 million years ago probably obliterated the dinosaurs.

3-SECOND BANG

If you have seen a shooting star, you have witnessed a meteor: the fiery final plunge of a chunk of space debris into Earth's atmosphere.

3-MINUTE ORBIT

Meteorites found on Earth, typically from bodies initially 1–10 metres (3–30 ft) across, are among the few tangible ambassadors that we have from outer space. Other astronomical objects are studied by the light they reflect or emit, but meteorites can be dissected in exquisite detail using sophisticated equipment. For example, radioactive dating of certain types of meteorites place their age, and that of our solar system, at precisely 4.56 billion years.

RELATED TOPICS

See also
THE EARTH
page 18

ASTEROIDS
page 42

COMETS
page 46

3-SECOND BIOGRAPHIES

LUIS & WALTER ALVAREZ
1911–88 & 1940–
American scientists who found evidence that a massive meteor impact could have caused the extinction of the dinosaurs, and other species

30-SECOND TEXT

Zachory K. Berta

When the Earth crosses through the eccentric, rubble-strewn orbits of comets, we experience more meteors than usual, a repeating event known as a meteor shower.

THE STARS

THE STARS
GLOSSARY

Algol Eclipsing binary star (pair of stars) in the constellation Perseus. Once every 69 hours, one star eclipses the other for about 10 hours. This means that its light appears to dip, noticeably enough to be seen by the naked eye. In many cultures, the star is associated with evil: *Al Gol* is Arabic for 'the demon', Hebrew tradition calls the star 'Satan's head' and the ancient Greeks saw it as the winking eye of a Gorgon (female monster) held by the hero Perseus.

black hole Region in which matter has been highly compressed, and as a result gravity acts with such force that everything in the area, even light, is drawn powerfully in. Black holes can come into existence when a massive star is dying.

blue giant star The most massive and hottest known type of star; it gives off bright blue light, often seen in regions of spiral galaxies where stars are being born.

Butterfly Nebula Also known as NGC 6302 and situated about 3,800 light-years distant in the constellation Scorpius, a planetary nebula given its descriptive name because its vast gas clouds resemble the wings of a butterfly. The clouds of gas, ejected by a dying star from a binary star system, have been made to glow by ultraviolet radiation also emitted by the star.

gamma ray burst Flash of high-frequency electromagnetic radiation, typically released during a supernova.

giant star One with a significantly larger luminosity and radius than a main sequence star, typically as much as 1,000 times as luminous as our Sun and with 10–100 times our Sun's radius. Even larger, more massive and more luminous stars are labelled 'supergiants' and 'hypergiants'.

main sequence stars Those on the main sequence of the Hertzsprung-Russell diagram of colour and brightness.

Mira Also known as Omicron Ceti, red giant star between 200 and 400 light-years away in the constellation Cetus. Mira is an example of a pulsating variable star. Its brightness varies on a regular cycle 332 days in length.

nebula (pl. nebulae) A cloud of dust or gas in interstellar space.

neutron star Extremely dense star created from exhausted nuclear fuel following the final explosion (supernova) of a massive star at the end of its life.

nova explosion Explosion in a white dwarf star caused when the star takes on matter from a twin in a binary star system and reignites, leading to runaway nuclear fusion on the surface of the white dwarf. The explosion is less powerful and less bright than a supernova. The name comes from the Latin word for 'new', because previously invisible white dwarf stars reappear when a nova occurs and may be taken for a new star.

planetary nebula A cloud of gas expelled into space by a red giant star. The term is derived from the German-born British astronomer William Herschel who, when he identified the phenomenon in 1785, thought the nebulae or clouds he viewed were similar to the 'gas giant' planet Uranus. Astronomers still use the term, although these nebulae form around dying stars and have no connection to planets.

red giant star A cooler star, of lower mass compared to a blue giant star.

Ring Nebula Also known as Messier 57, a planetary nebula in the constellation of Lyra. It consists of a cloud of ionized gas expelled into space by a red giant star.

supernova (pl. supernovae) Explosion at the end of a star's life, when the core of a massive star collapses to form a black hole or a neutron star. A particular type of supernova, type 1a, occurs when a white dwarf star takes in material from a companion in a binary star system until, passing a critical mass (1.4 times that of our Sun), it explodes.

supernova remnant Structure created by a supernova explosion, containing the material of the star that exploded and any interstellar material swept along with it.

white dwarf star Highly dense remnant of a star, created after a red giant star swells and gives issue to a vast nebula, exposing the star's core – which cools and grows dim to form a white dwarf.

COLOUR & BRIGHTNESS OF STARS

the 30-second astronomy

We perceive colour in stars when they have an uneven spectrum of light. Some emit more blue light than red, some the reverse – like the colour of hot irons in a fire, the stars' colour indicates their surface temperature, with blue stars being hottest (20,000°C/36,000°F) and red stars coolest (3,000°C/5,500°F or less). Astronomers code colour in most stars with a sequence of seven letters – O, B, A, F, G, K and M, running from hot to cool; stars are also ranked by brightness – supergiants are very bright, giants less bright, and dwarfs less bright still. The Sun is middle-ranking: a G-type dwarf. In about 1910, astronomers Ejnar Hertzsprung and Henry Russell, each working independently, plotted the brightness of a number of stars against their temperature, creating the Hertzsprung-Russell (H-R) diagram. Brightness and colour are properties of a star's surface, but the H-R diagram reveals what is happening inside. Most stars sit on the main sequence from blue/bright to red/faint – and a star's mass determines where it sits, with the most massive stars at the bright end and the least massive ones at the faint end.

3-SECOND BANG
The surface brightness and colour of stars is plotted on the Hertzsprung-Russell diagram – the key to how stars live and die.

3-MINUTE ORBIT
Stars on the main sequence are changing their hydrogen into helium, releasing nuclear energy. As a main-sequence star uses up the hydrogen in its core, it builds a shell of helium around the core, and uses this as its next source of nuclear fuel. The star grows brighter, but also cooler, and becomes a giant star, even a supergiant. Supergiants finally explode, but giants contract again, becoming dimmer white dwarfs, fading gently to blackness.

RELATED TOPICS
See also
GIANT STARS
page 60

WHITE DWARFS
page 62

SUPERNOVAE
page 68

3-SECOND BIOGRAPHIES
EJNAR HERTZSPRUNG
1873–1967
Danish astronomer

HENRY RUSSELL
1877–1957
American astrophysicist

30-SECOND TEXT
Paul Murdin

The main sequence of dwarf stars runs diagonally across the Hertzsprung-Russell diagram; white dwarfs occupy the lower left corner and super giants the upper right.

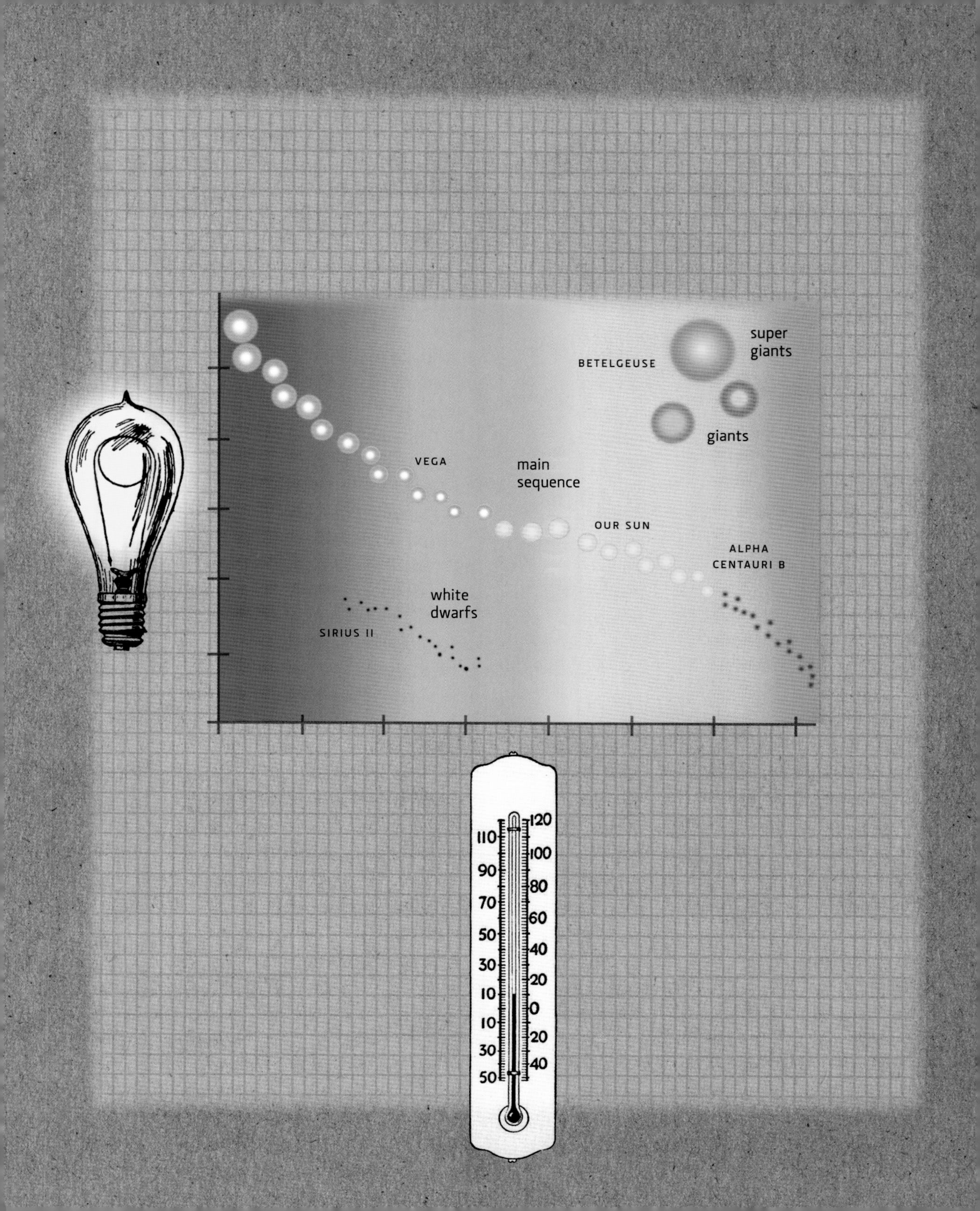

super giants
BETELGEUSE
giants
VEGA
main sequence
OUR SUN
ALPHA CENTAURI B
white dwarfs
SIRIUS II
120
110
100
90
80
70
60
50
40
30
20
10
0
10
20
30
40
50

BINARY STARS

the 30-second astronomy

When stars form from massive gas clouds, there is often enough gas to make two stars. Astronomers estimate that about half the stars we see are actually pairs of stars orbiting one another – binary stars. If the planet Jupiter had been born 100 times more massive, then along with the Sun, it, too, would have been a star, and we would be living in a binary star system. There are many types of binary star, because the two stars involved can differ greatly, depending on their mass at birth. Massive stars live fast and die young, becoming black holes, neutron stars or white dwarf stars, while its companion is still the stellar equivalent of a teenager. Sometimes binary stars are so close that one star can strip material off its companion. Other binaries are more peaceful. Some eclipse, where one star hides its companion as they orbit each other. As the companion reappears, unique clues as to the makeup of these systems are revealed. One of the best-known binary stars is Algol. Every 69 hours, it fades by a factor of 3 for almost 10 hours, while the fainter member of the pair hides the brighter star.

3-SECOND BANG
Stars often form in pairs, so if you gaze upon a lonely-looking star, for about half the time it will have a fainter, invisible companion.

3-MINUTE ORBIT
Astronomers can learn a huge amount from binary stars. By watching how quickly the stars orbit one another, we can accurately determine the masses of those two stars, and so establish the mass of all similar stars. Astronomers have also seen stars orbiting black holes in binary star systems, and watching the speed of the orbiting star is the best evidence we have for the existence of black holes.

RELATED TOPICS
BLACK HOLES
page 70

COSMIC X-RAYS
page 104

3-SECOND BIOGRAPHIES
WILLIAM HERSCHEL
1738–1822
German-born British astronomer who coined the term 'binary star' in 1802

ÉDOUARD ROCHE
1820–83
French astronomer and mathematician who calculated how binary stars could affect each other

30-SECOND TEXT
Darren Baskill

This artist's impression of a binary system shows the stars so close together that gas is flowing from the sunlike star onto its smaller, white dwarf companion.

VARIABLE STARS

the 30-second astronomy

Variable stars vary in brightness in different ways and for many reasons. Pulsating ones vary in size and brightness in a fairly predictable and regular way. Gravity makes such stars shrink, causing an outer layer of helium to block light escaping from beneath; the energy of this blocked light is then absorbed by the helium, causing the gas and the entire star to expand again. Once it has expanded to a large size, the helium becomes transparent again, and the heat can escape into space, causing the star to cool and collapse – and the cycle repeats itself. The star Mira (in the constellation Cetus) is an example of a pulsating variable star: its regular variation was discovered in 1638. Every 332 days, Mira varies from being visible with the unaided eye to requiring a telescope to see it. Cataclysmic variable stars vary dramatically and unpredictably. These include: dwarf novae, in which a massive avalanche of gas falls through a disc surrounding a star as frequently as every few weeks; novae, in which the outer surface of a white dwarf star suddenly explodes; and supernovae, in which the whole of either a white dwarf or massive star explodes.

3-SECOND BANG

Most stars vary in brightness. Some alter to a degree that is hardly noticeable, others – which we call variable stars – change significantly.

3-MINUTE ORBIT

Variable stars are an active area of research involving professionals and amateurs working together. While professional astronomers scrutinize individual stars, amateurs can scan the entire sky for new or unusual behaviour. Amateurs who spot a variable star behaving unusually can contact variable star organizations, whose members inform professional astronomers. Within hours, the largest telescopes on Earth – or even space telescopes – can follow up an amateur's observation to see in detail what that star is doing.

RELATED TOPICS

See also
WHITE DWARFS
page 62

SUPERNOVAE
page 68

3-SECOND BIOGRAPHIES

JOHANNES HOLWARDA
1618–51
Frisian astronomer who discovered in 1638 that Mira was a variable star

30-SECOND TEXT

Darren Baskill

Astronomers commonly observe a rise and fall in the brightness of stars over hours, decades, or even longer. One, SCP O6F6, discovered by the Hubble telescope in 2006, grew steadily brighter for 100 days, then dimmed back to oblivion after another 100 days.

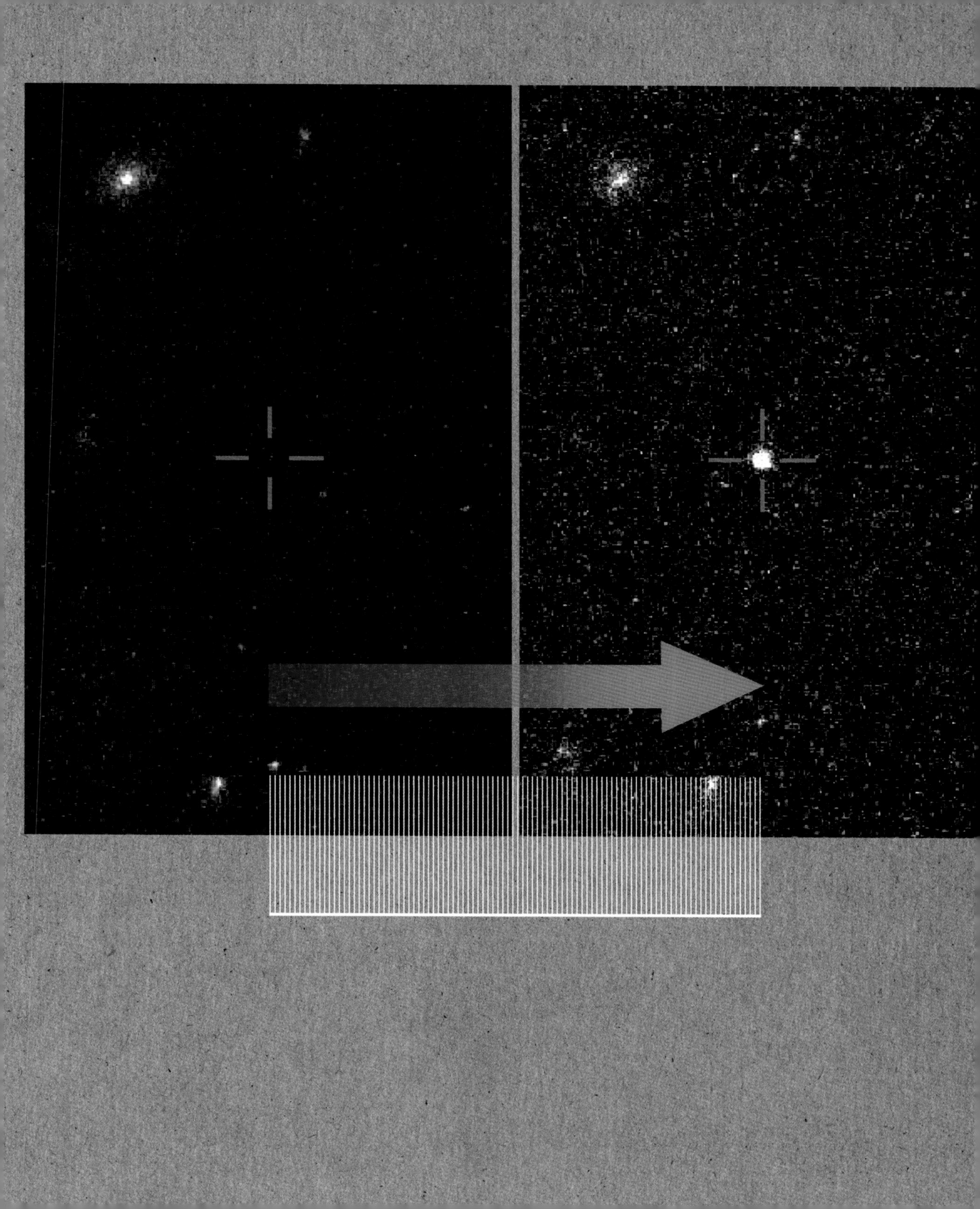

GIANT STARS

the 30-second astronomy

3-SECOND BANG
Giant stars are those that are between 10 and 100 times larger than the Sun, and up to 1,000 times brighter.

3-MINUTE ORBIT
Stars that are even larger and brighter are known as supergiants and hypergiants. The current record holder for the largest known star is held by VY Canis Majoris, a red hypergiant about 2,000 times larger than our Sun, and 500,000 times more luminous. If placed at the centre of the solar system, VY Canis Majoris's surface would extend beyond the orbit of Jupiter.

The wide range in colour and brightness of stars arrayed in the Hertzsprung-Russell diagram demonstrates that there are many with ages, sizes, luminosities and masses very different from those of our Sun. The very rarest kind are the giant stars. Blue giants are the hottest and most massive stars. They produce energy at prodigious rates to withstand the inward pull of gravity, and thus consume the available fuel very rapidly to have lifetimes lasting only a few million years. Their characteristic bright blue light dominates the open clusters of stars that trace the recent star formation arms of spiral galaxies. Red giants, in contrast, are more common. They are lower-mass stars that have evolved beyond creating energy through simple fusion of hydrogen to helium at the centre. In a red giant, the core begins to compress under gravity and heats up to trigger subsequent, and more complicated, phases of nuclear burning. The resulting increase in luminosity inflates the giant's outer layers, so that the surface of the bloated star is then at a cooler temperature, and appears much redder in colour. Such stars end their lives when they blast themselves apart as a planetary nebula or a supernova.

RELATED TOPICS
See also
COLOUR & BRIGHTNESS OF STARS
page 54

SUPERNOVAE
page 68

3-SECOND BIOGRAPHIES
EJNAR HERTZSPRUNG
1873–1967
Danish astronomer

HENRY RUSSELL
1877–1957
American astrophysicist

30-SECOND TEXT
Andy Fabian

The term 'giant star' is no mere hyperbole: Betelgeuse, for example, is an evolved star with a radius about 1,200 times greater than that of the Sun.

WHITE DWARFS

the 30-second astronomy

3-SECOND BANG
White dwarfs are the cinders of dead stars – although abundant, they are faint or even invisible, and therefore difficult to find in interstellar space.

3-MINUTE ORBIT
An unexpected feature of the theory of white dwarfs is that electron degeneracy pressure is only effective in propping up a white dwarf star if its mass is less than 1.4 times the mass of the Sun. When the core of a star with more mass than this tries to form a white dwarf, it collapses and becomes a black hole.

Stars, such as the Sun, eventually become giant stars and swell up; this gives them a reduced surface gravity and their outer layers escape, forming a nebula. Such a nebula often has a beautiful circular or bilateral symmetry, sometimes expressed in its name; examples are the Ring Nebula and the Butterfly Nebula. As the nebula forms, the hot inner core of the red giant star becomes exposed at its centre, energizing the nebula and giving it wonderful colours. The naked core is spent nuclear fuel; inside the star it has been heated to become very hot indeed, and it radiates powerfully. Now exposed, it cannot long stay hot. It quickly cools and dims, while the nebula fades away and dissipates. The star becomes an isolated 'white dwarf' – faint, small (Earth-sized), and dense, a cooling, inert stellar cinder, eventually completely dark. White dwarf stars are so dense that they have a strong gravitational field that works to make them collapse. Holding white dwarf stars up is a kind of pressure revealed by quantum mechanics, unknown before 1925, called 'electron degeneracy pressure'. It is surprising that a phenomenon of the really small is needed to prop up a star.

RELATED TOPICS
See also
PULSARS
page 64

BLACK HOLES
page 70

MOLECULAR CLOUDS & NEBULAE
page 78

MESSIER OBJECTS
page 80

3-SECOND BIOGRAPHY
SUBRAHMANYAN CHANDRASEKHAR
1910–95
Indian-American astrophysicist who studied white dwarfs

30-SECOND TEXT
Paul Murdin

A red giant star loses its outer layers, which surround the star with a beautiful nebula. The star becomes a small, white dwarf.

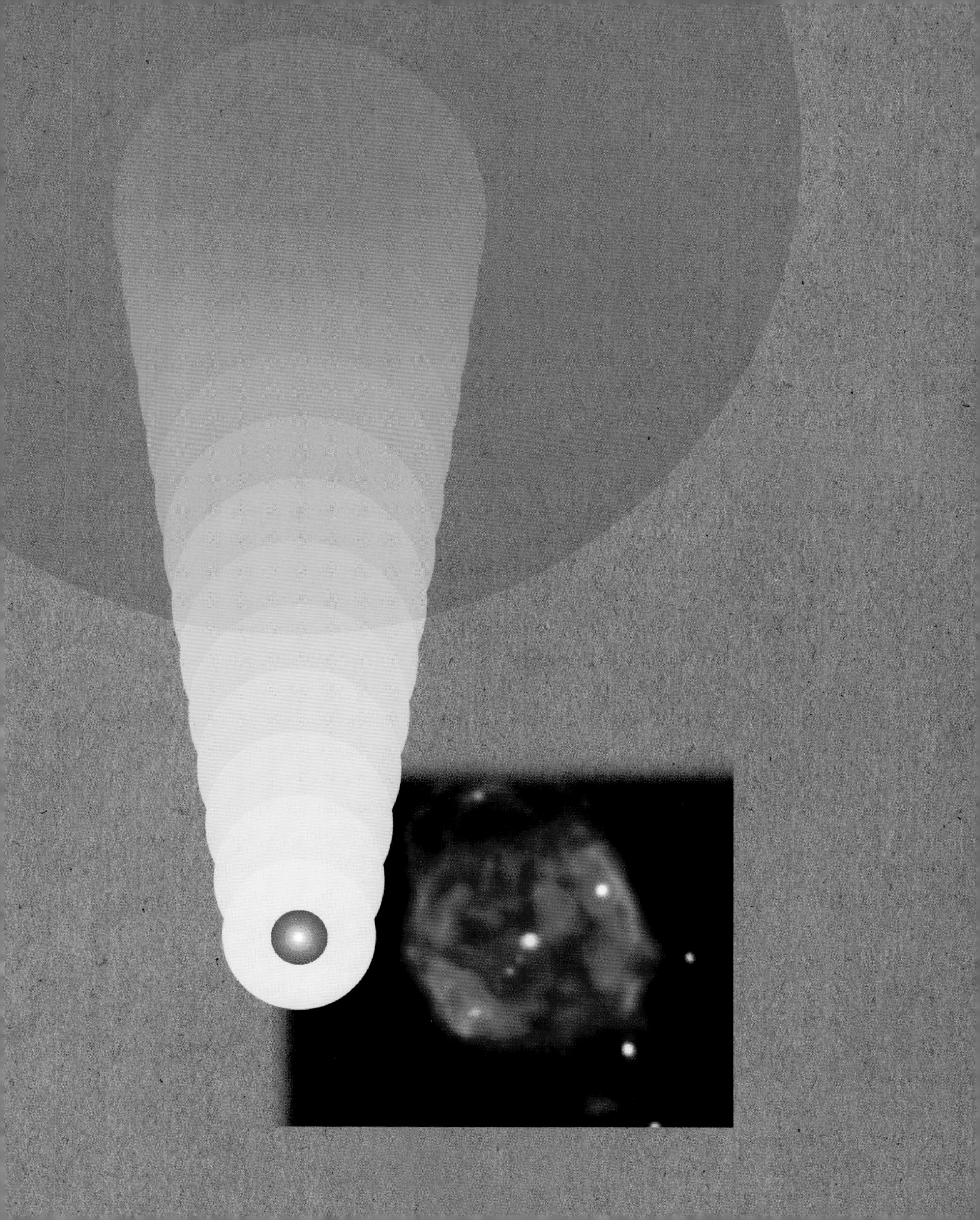

PULSARS

the 30-second astronomy

Like white dwarfs, neutron stars are stellar cinders, created from the cores of exhausted nuclear fuel inside some massive stars when, at the end of their lives, they explode as supernovae. The core collapses to an extremely dense star made of neutrons, about the mass of the Sun but typically only 15–25 kilometres (10–15 miles) in diameter; the density is comparable to that of a mountain compressed so small it will fit into a teaspoon. During the collapse, the rotation of the core is greatly speeded up, much as the spinning of an ice skater speeds up as she draws her outstretched arms to her sides. The star may have been rotating once every day or month; compressed to a neutron star, the core rotates faster than once every second. At the same time, any magnetic field that was threaded through the core is greatly intensified. The magnetic field generates a broad spectrum of radiation, including radio radiation, that beams into space. If the star's rotation sweeps the beam towards the direction of Earth, the star appears to us to pulse, like a lighthouse. The phrase 'pulsating radio star', used to describe neutron stars, was contracted to 'pulsar'.

3-SECOND BANG
Formed in supernova explosions, pulsars are neutron stars that reveal their presence as pulsating radio stars.

3-MINUTE ORBIT
Some pulsars exist in binary stars; some of these are a neutron star in orbit around a more usual star; some of them two neutron stars. Binary neutron stars lose energy and approach one other, and – although this has not been observed – are thought eventually to merge with a huge explosion, causing a gamma ray burst and a black hole.

RELATED TOPICS
See also
SUPERNOVAE
page 68

BLACK HOLES
page 70

GAMMA RAY BURSTS
page 106

THE LIGHT SPECTRUM
page 122

3-SECOND BIOGRAPHIES

ANTHONY HEWISH
1924–
British radio astronomer, Bell Burnell's supervisor, and codiscoverer of pulsars

JOCELYN BELL BURNELL
1943–
British astronomer, the discoverer of pulsars

30-SECOND TEXT
Paul Murdin

A dense, small pulsar energizes the swirling central gases of the Crab Nebula – a supernova remnant and pulsar wind nebula.

15 July 1943
Born in Belfast

1954
Attended a Quaker school in York

1965
Graduated in physics from University of Glasgow

1967
First observation of what will become known as the first pulsar, CP 1919

1968
First use of the word 'pulsar'

1969
Completed PhD at Cambridge University

1974
Anthony Hewish and Martin Ryle shared the Nobel Prize for Physics; Bell Burnell was not cited

1978
Won Robert Oppenheimer Memorial Prize

1979
Published 'Little Green Men, White Dwarfs or Pulsars?' in *Cosmic Search Magazine*

1987
Won Beatrice M. Tinsley Prize, American Astronomical Society

1989
Awarded Herschel Medal, Royal Astronomical Society, London

1991
Made Professor of Physics at the Open University, and Visiting Professor at Princeton

1999
Awarded British order of chivalry, CBE (Commander of the Order of the British Empire), for services to astronomy

2001–04
Dean of Science at University of Bath

2002–04
President of Royal Astronomical Society

2003
Fellow of the Royal Society

2008
Created Dame

2008–10
The first-ever female president of the Institute of Physics

JOCELYN BELL BURNELL

Susan Jocelyn Bell was born in 1943 in Belfast, Northern Ireland, into a Quaker household. At the age of 11, she failed to gain the necessary qualifications for entry to a grammar school, a type of publicly run British school reserved for more academically able children. This failure was a setback, and she has said that her determination to overcome it spurred her on through her career in astronomy, at the time a difficult and lonely course for a woman.

While studying for her PhD at Cambridge under the supervision of Anthony Hewish she made the discovery that changed the way we think of the Universe. Bell's field is radio astronomy: as part of her PhD, she helped to build and maintain the Interplanetary Scintillation Array, a huge (1.6-ha/4-acre) radio telescope at the Mullard Radio Astronomy Observatory, Cambridge. Part of Bell's work was to interpret the mass of data that the telescope generated every 24 hours, and in November 1967 she noticed what has become famous as a 'scruff' on the graph. This was a tiny detail and could have easily been missed, but Bell followed it up. It was finally identified as a hitherto unobserved phenomenon, a pulsating star ('pulsar'), later named CP 1919. The discovery thrilled the astronomical world, and there was jocular talk of 'little green men', because one of the theoretical explanations of the regular pulsating radio waves was that there might be 'somebody out there' signalling into the void. Eventually, it was determined that pulsars were neutron stars that emitted regular waves of radiation. Bell – known as Jocelyn Bell Burnell after her marriage to Martin Burnell in 1968 – discovered three more pulsars, opening up a whole new field in astrophysics.

Although Bell Burnell's name is now indissolubly linked with this breakthrough discovery, it caused a controversy beyond the world of astrophysics that is still unresolved. Bell Burnell's name was second on the paper announcing the discovery, but the Nobel Prize that it won went to Hewish and Martin Ryle (leader of the research team), with no mention of Bell Burnell. It is not usual for research assistants to go uncredited, despite their doing most of the grunt work; however, many considered that pulsars had been recognized only through Bell Burnell's persistence and attention to detail, and her omission created an outcry – with Sir Fred Hoyle, in particular, championing her cause.

SUPERNOVAE

the 30-second astronomy

During the late stages of its evolution, a massive star produces energy through the manufacture of progressively heavier elements deep in its core. The end point of this process for stars above 8 Solar masses occurs at the creation of iron, beyond which energy is no longer released by fusion. The star then abruptly runs out of fuel. The core implodes under gravity to form either a neutron star or a black hole; as it contracts, it becomes denser and hotter, releasing so much energy that the supernova explosion can temporarily outshine its host galaxy. The rest of the star is ripped apart, and hot debris is propelled outwards by the blast to form a shell of material expanding away at some 15,000 m/s (16,000 ft/s), sweeping up any interstellar gas before it. The material is compressed into a filamentary structure known as a supernova remnant. An intense flood of neutrons released in the explosion allows the creation of the very heaviest elements. Both these, and those originally forged at the core of the star, are blasted out to mix with surrounding clouds of gas and dust, where they will eventually be recycled into new generations of stars and planets.

3-SECOND BANG
The end of a massive star's life is heralded by one of the largest explosions in the Universe, known as a supernova.

3-MINUTE ORBIT
A separate type of supernova explosion can be triggered if a white dwarf steadily accretes matter from a large companion star in a binary system. Once its total mass crosses the threshold of 1.4 Solar masses, it will explode completely to form a supernova Type 1a. Such supernovae are, however, rare events: it is estimated that one occurs in an average galaxy only about once a century.

RELATED TOPICS
See also
BINARY STARS
page 56

GIANT STARS
page 60

WHITE DWARFS
page 62

3-SECOND BIOGRAPHIES
WILLIAM FOWLER
1911–95
American astrophysicist

FRED HOYLE
1915–2001
British astronomer

MARGARET & GEOFFREY BURBIDGE
1919– & 1925–2010
British-American astronomers

30-SECOND TEXT
Andy Fabian

A giant star ends its life in a supernova explosion, leaving a neutron star or black hole surrounded by a rapidly expanding shell of hot gas.

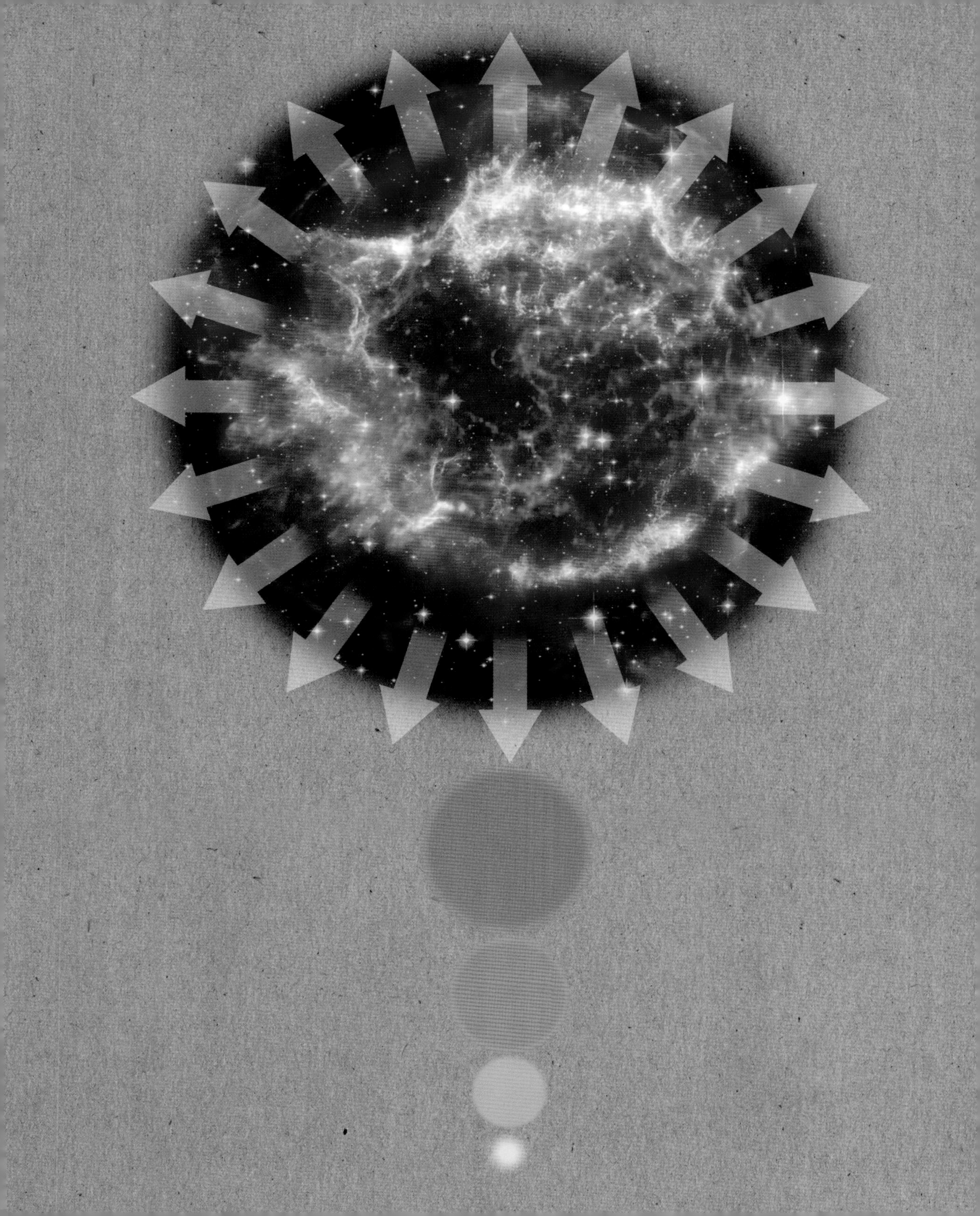

BLACK HOLES

the 30-second astronomy

The existence of black holes was first suggested by eighteenth-century English philosopher John Michell, who in 1783 wondered if stars could exist with a mass so large and a gravity so strong that nothing, not even light, could escape them. He called them dark stars, which aptly describes black holes. We now know that they come in many sizes. Stellar black holes squeeze the mass of ten Suns into an area the size of London, and we know of dozens within our Milky Way Galaxy; supermassive black holes have a mass between 1 million and 10 billion times that of our Sun, and are found at the heart of many galaxies, including our own; and a few intermediate black holes are known with masses in between these extremes. While black holes have a reputation for immensely strong gravity, their effects are felt only at close quarters. If an object is close to a black hole, its nearer parts feel so much more gravity than its further parts that it is stretched into long, thin strands – a process known as 'spaghettification'. Fortunately, the nearest black hole to us lies more than 3,000 light-years away.

3-SECOND BANG
A black hole is a region in which matter has been extremely compressed, making gravity so powerful that everything in the vicinity is sucked in.

3-MINUTE ORBIT
Looking for dark stars against the blackness of space is extremely difficult. Astronomers do not look for black holes directly, but for their effects on their surroundings. When dismembered stars fall in towards a black hole, they are heated so much that they emit X-rays; by seeing how quickly a star orbits a black hole, astronomers can calculate the mass of both – clues that indicate the presence of black holes.

RELATED TOPICS
See also
THE MILKY WAY
page 82

GAMMA RAY BURSTS
page 106

QUASARS
page 108

3-SECOND BIOGRAPHIES
JOHN MICHELL
1724–93
English philosopher who first proposed that black holes exist

KARL SCHWARZSCHILD
1873–1916
German physicist who solved Einstein's General Relativity equations to describe conditions around a black hole

30-SECOND TEXT
Darren Baskill

A star that wanders too close to a black hole is spaghettified – vertically stretched and horizontally compressed – due to the hole's very strong gravitational field.

THE MILKY WAY

THE MILKY WAY
GLOSSARY

Andromeda Galaxy Also known as Messier 31, the nearest galaxy to our own (the Milky Way), apart from a number of smaller 'companion galaxies' that are satellites of the Milky Way. The Andromeda Galaxy is a spiral galaxy situated about two-and-a-half-million light-years away in the Andromeda constellation and contains 1 trillion (1,000,000,000,000) stars.

Cepheid variable Type of star that varies between a compressed and an expanded state. The stars (which have a mass of between 5 and 20 times that of our Sun) expand when pressure builds up, then contract because pressure is lower in the star's expanded state. Cepheids are used by astronomers to establish extragalactic distances.

comet An icy body with a coma (temporary atmosphere) and a tail in orbit around the Sun that becomes visible when sufficiently close to the Sun. The tail points away from the Sun, while the curved coma follows the comet's path.

diffuse star formation nebulae Higher density clouds in the interstellar medium, in which stars are formed.

elliptical galaxy Galaxy in the form of an ellipsoid (a three-dimensional ellipse). One of three galaxy types identified in 1936 by American astronomer Edwin Hubble, along with lenticular and spiral galaxies.

galaxy A system bound together by gravity, containing stars, clouds of dust and gas in the interstellar medium and dark matter.

globular cluster Group of stars, tightly bound in a spherical shape by gravity, that orbits the core of a galaxy. There are between 150 and 160 globular clusters in the Milky Way, containing some of the oldest stars in the Galaxy.

interstellar medium The matter that fills the space between the star systems within a galaxy. Also known as ISM, it consists principally of gas and dust; from these materials, new stars are formed, and light generated by their birth heats the remaining gas atoms and makes pink-red nebulae visible in the vicinity of the new stars.

lenticular galaxy Type of disc galaxy with a central group of stars (called a bulge), similar to a spiral galaxy, but lacking the spiral arms of the spiral galaxy shape.

nebula (pl. nebulae) Visible cloud of dust or gas in interstellar space. Emission nebulae are visible because gas atoms within them have been heated by ultraviolet light emanating from a nearby star; reflection nebulae can be seen because they are reflecting light from a local star or star group; dark nebulae are identifiable because they block the light from a star or star group behind them.

open cluster of stars Loosely bound group of stars held together by mutual gravitational attraction and in orbit around the centre of a galaxy. The Pleiades group is an example of an open cluster. There are believed to be more than 1,100 open star clusters within the Milky Way Galaxy.

Orion Nebula Also known as Messier 42, a vast region of gas and dust – 13 light-years across – situated just to the south of Orion's belt in the constellation of Orion.

Pleiades cluster A group of stars, also known as Messier 45, another of the 'Messier objects' identified by the eighteenth-century French astronomer Charles Messier. The cluster is about 425 light-years' distance from the Earth, and it contains hundreds of stars, although only a few are visible to the naked eye. These bright blue stars are also known as 'the Seven Sisters'.

spiral galaxy Type of galaxy with a central group of stars (a bulge), and with spiral arms containing stars, gas and dust, extending outwards into a rotating disc-like structure.

standard candle Astrophysical object with known luminosity that can be used to calculate the distance from Earth of a formation in which it appears. Cepheid variables are used as standard candles.

starburst Period of intense star formation. In a starburst phase, stars form at a rate of up to 100 times that of regular stars.

Triangulum Galaxy Also known as Messier 33 or the 'Pinwheel Galaxy', a spiral galaxy in the constellation Triangulum. Part of the Local Group of galaxies that includes the Milky Way and the Andromeda Galaxy, the Triangulum lies 3 million light-years from Earth but can be seen with the naked eye in very good conditions, making it one of the most distant objects that can be viewed without a telescope.

Virgo cluster of galaxies Group of galaxies in the constellation Virgo. The group contains up to 2,000 galaxies and is at the heart of an even larger Virgo supercluster. The Local Group of galaxies, which contains our galaxy (the Milky Way) and the Andromeda Galaxy, belongs to the Virgo supercluster and is orbiting around the Virgo cluster.

CONSTELLATIONS

the 30-second astronomy

The different stars we can see in each constellation are very rarely physically associated with each other, but just happen to be in the same line of sight from the Earth. Constellations are grid-like segments around recognizable patterns of stars called 'asterisms'. For example, the asterism known as the 'Big Dipper' corresponds to the seven brightest stars in the larger Ursa Major constellation. The majority of the constellations take their names from the astronomical treatise *Almagest,* written in the second century AD by Egyptian astronomer Claudius Ptolemy. The apparent motion of stars in the sky is due to the Earth's rotation on its axis. We see them rotating around the Earth's rotation axis, which is practically aligned with the North (or Pole) Star – Polaris. The planets and the Moon all orbit the Sun on roughly the same plane, which is named the ecliptic. Consequently, the Sun, the other planets and the Moon all appear as moving objects on the same circle in the Earth's celestial sphere, in a background of stars that scarcely move during a human lifetime. The constellations crossed by this circle are the 13 'zodiacal' constellations.

3-SECOND BANG

The celestial sphere is divided into 88 constellations – regions of the sky around recognizable patterns of stars as perceived from the Earth.

3-MINUTE ORBIT

Billions of people believe that human behaviour is influenced by the position of planets in the sky, the phase of the Moon, or the 'zodiacal sign' (based on the constellation in which the Sun used to rise on their birth date some 2,000 years ago). However, there is not the least clue that any physical property related to astronomical objects (such as gravity or light) could have an influence on people.

RELATED TOPICS

See also
THE MOON
page 20

COLOUR & BRIGHTNESS OF STARS
page 54

THE MILKY WAY
page 82

3-SECOND BIOGRAPHIES

CLAUDIUS PTOLEMY
c. AD 100–c. AD 170
Egyptian astronomer

NICOLAS LOUIS DE LACAILLE
1713–62
French astronomer who catalogued 10,000 stars

30-SECOND TEXT

François Fressin

The seven stars that form the 'big dipper' in the constellation Ursa Major are not related, and all are at different distances from the Earth.

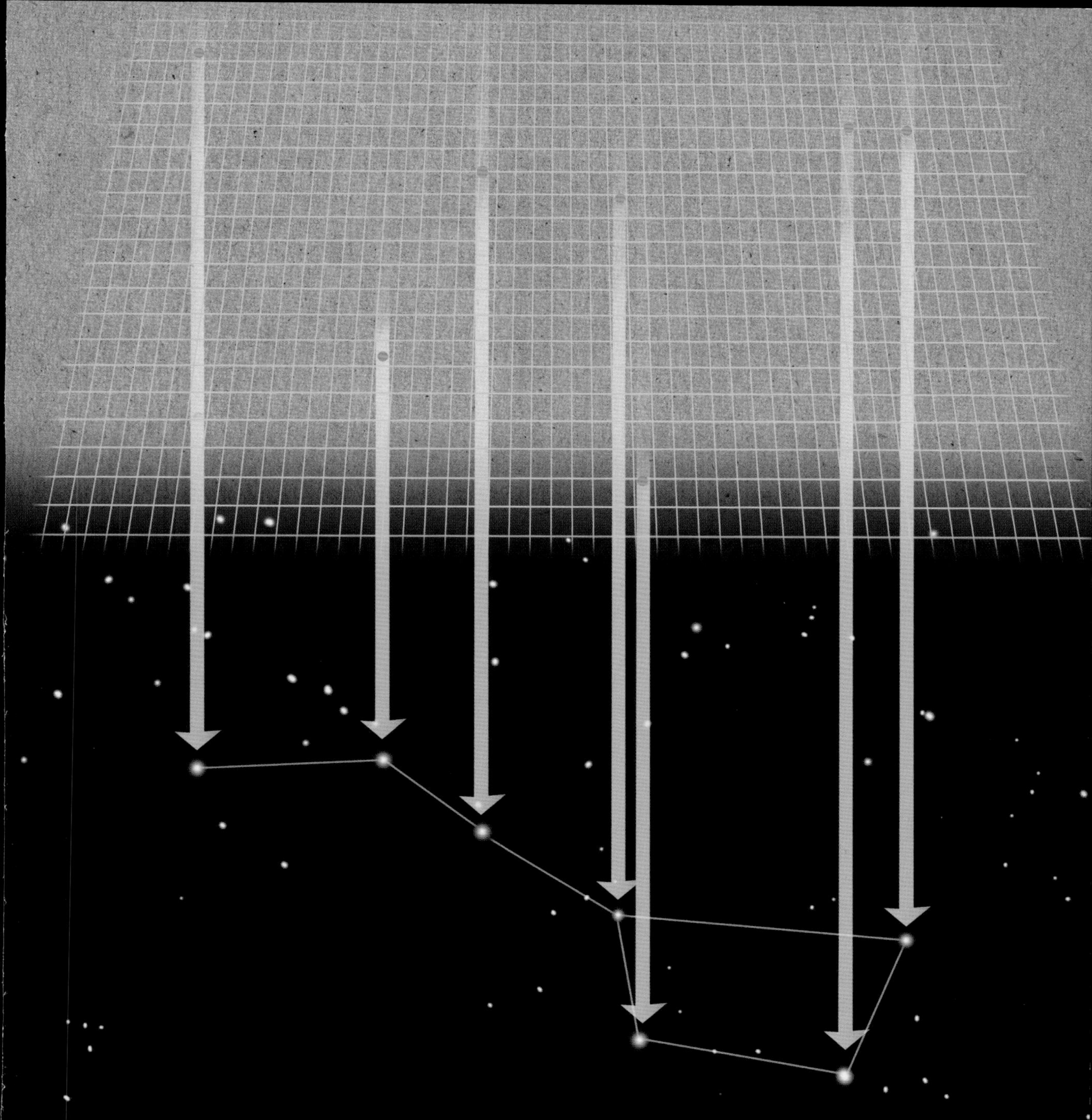

MOLECULAR CLOUDS & NEBULAE

the 30-second astronomy

3-SECOND BANG
Interstellar gas clouds are the fundamental reservoir from which new stars form, along with their planetary systems – and any life forms living on those planets.

3-MINUTE ORBIT
Tiny solid particles ('dust grains') are mixed with the gas. The densest concentrations within a nebula form opaque silhouettes against background light. The dust protects the core of such clouds from heat and light; temperatures plummet to a few degrees above absolute zero, and the atoms can form complex molecules. These molecular clouds have typical sizes between 3 and 50 light-years, and can contain up to 1,000 Solar masses of matter.

The space between the stars is not completely empty but teems with atoms and molecules of gas, making up the interstellar medium. The combined mass of all the gas in the Milky Way is only one-tenth of that contained in stars, yet it is clumped into diffuse clouds that extend out way beyond the spiral arms of a galaxy. At temperatures of tens to hundreds of degrees above absolute zero, the matter in the clouds is so cold that it is mostly in the form of neutral hydrogen atoms, and is thus transparent at visible wavelengths. Traces of heavier elements, such as carbon, oxygen and iron, pepper the clouds, which have been recycled from the hearts of massive stars by the explosions that mark the end of the stars' lives. The coldest, densest pockets within these clouds provide the ideal conditions for starbirth. Stars newly formed from the interstellar medium flood the surrounding clouds with energetic light, heating the gas atoms and causing them to glow. The gas then becomes apparent as the distinctive, pinky-red nebulae that accompany young blue star clusters lining the spiral arms of a galaxy.

RELATED TOPICS
See also
SUPERNOVAE
page 68

MESSIER OBJECTS
page 80

THE MILKY WAY
page 82

GALACTIC STRUCTURES
page 88

3-SECOND BIOGRAPHY
BART BOK
1906–83
Dutch-born American astronomer

30-SECOND TEXT
Carolin Crawford

The gas atoms in a diffuse cold cloud of interstellar gas are heated in the vicinity of newly formed clusters of stars; they then emit visible light to glow as nebulae.

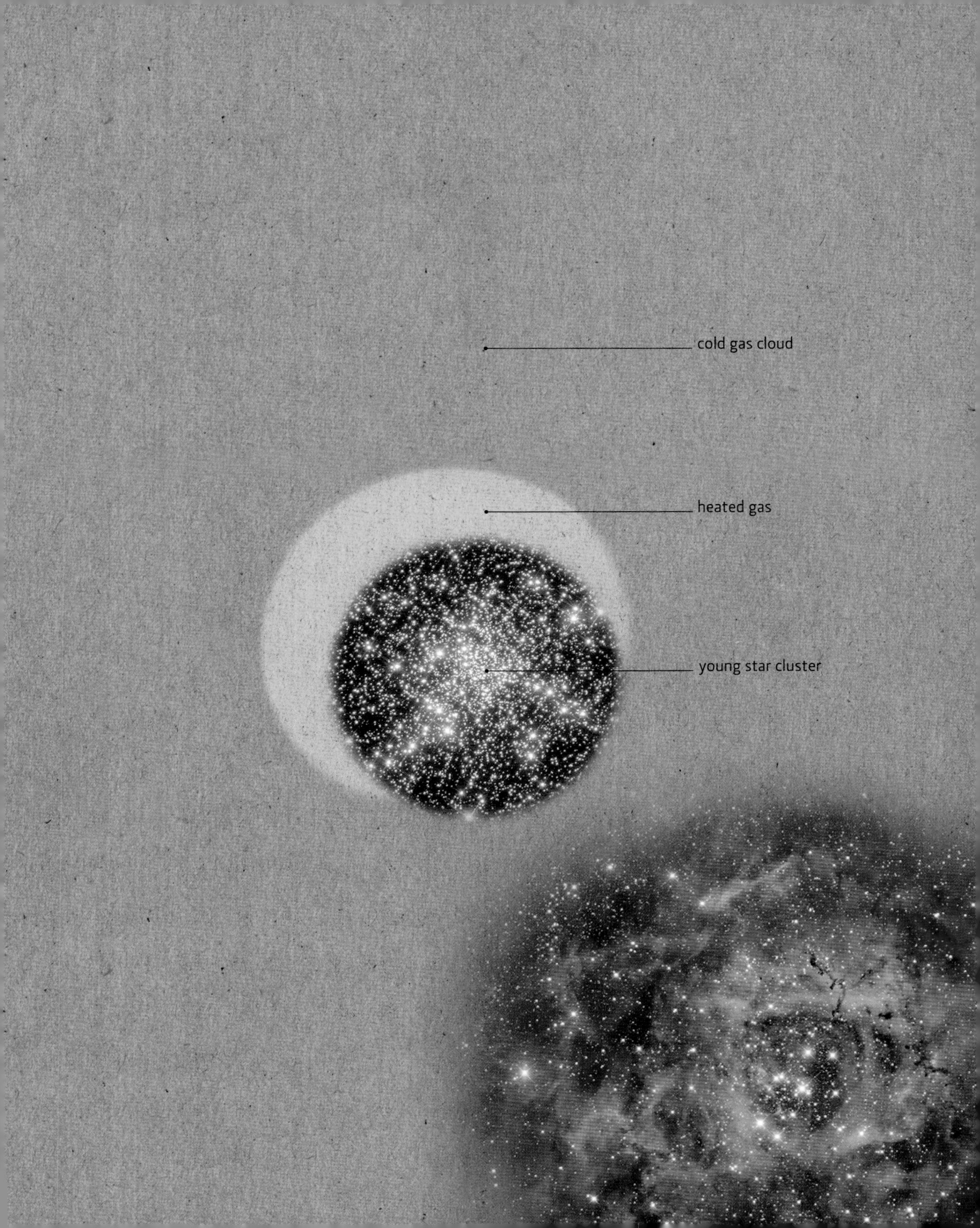
cold gas cloud
heated gas
young star cluster

MESSIER OBJECTS

the 30-second astronomy

Charles Messier was one of the first 'comet hunters', dedicated to the discovery and observation of new comets. These would appear through his telescope as indistinct, faint and fuzzy patches of light, whose nature was revealed only by their motion from day to day against the fixed backdrop of stars. Messier became frustrated in his searches by the presence of other faint structures, which – unlike the comets – were permanent features of the sky. To aid against confusion, he compiled a catalogue of such nebulae that could be mistaken for comets; many were original discoveries, although some were either visible to the unaided eye (such as the Orion Nebula, or the Pleiades cluster) or taken from the work of other astronomers, such as Edmond Halley. Ironically, Messier is remembered today less for the comets he did discover, and more for his catalogue of objects that were not comets. The modern Messier catalogue collects 110 diverse objects, from globular and open clusters of stars to supernovae remnants, planetary nebulae and diffuse star formation nebulae. The catalogue also includes 40 galaxies (identified as nebulae before their true nature was established), 16 of which comprise the nearby Virgo cluster of galaxies.

3-SECOND BANG

The eighteenth-century French astronomer Charles Messier created a catalogue of nebulae that lists some of the most interesting objects in the sky.

3-MINUTE ORBIT

Messier objects are often among the closest and best-known examples of their type. Messier's original observations were carried out with a telescope comparable in power to simple instruments available today, so his catalogue provides a selection of targets suitable for amateur astronomers. A 'Messier marathon' is an attempt to observe as many Messier objects as possible during one night; only during late spring and from low northern latitudes can a person observe them all.

RELATED TOPICS

See also
COMETS
page 46

MOLECULAR CLOUDS & NEBULAE
page 78

THE OTHER GALAXIES
page 86

GALACTIC STRUCTURES
page 88

3-SECOND BIOGRAPHIES

EDMOND HALLEY
1656–1742
English astronomer

CHARLES MESSIER
1730–1817
French astronomer

30-SECOND TEXT

Carolin Crawford

Comet hunter Charles Messier compiled a catalogue of some of the brightest, best-known astronomical objects in the sky.

THE MILKY WAY

the 30-second astronomy

All of the stars visible to the unaided eye are contained in our home galaxy, the Milky Way. They are flattened into a disc-like structure 100,000 light-years across, and their light merges to form a band of diffuse light that arches across the night sky. Clusters of blue stars, glowing nebulae, and dark rifts of star dust trace out the spiral arms. The Sun lies within this disc, halfway between the centre and its outer edge. A central bulge of older stars contains a dormant supermassive black hole at its core, with a mass 4 million times that of our Sun. Along with all the stars in the disc, the Sun rotates around the centre of the Galaxy in response to the gravitational pull of all the material closer in; travelling at 220 km/s (135 miles/s), it takes 240 million years to complete one orbit. The outer stars rotate too fast to remain bound to the Galaxy, but do not – as would be expected – spin off from their orbits. This suggests that there is much more matter providing gravitational attraction for the stars than is observed in the stars and gas, and is evidence for the presence of 'dark matter'.

3-SECOND BANG
The Sun is just one of 100,000 million stars that trace a vast spiral-shaped galaxy we call the Milky Way.

3-MINUTE ORBIT
The Milky Way is a member of the 'Local Group' of about 30 galaxies, along with the Andromeda and Triangulum spiral galaxies, and many dwarf galaxies in orbit around each of them. We are being steadily pulled by gravity towards our twin, the Andromeda Galaxy, which lies some two-and-a-half-million light-years away. The two galaxies are expected to merge in about 6 billion years' time, forming a new, much larger galaxy.

RELATED TOPICS
See also
MOLECULAR CLOUDS & NEBULAE
page 78

GALACTIC STRUCTURES
page 88

DARK MATTER
page 110

LIGHT-YEARS & PARSECS
page 118

3-SECOND BIOGRAPHIES
HEBER CURTIS
1872–1942
American astronomer

HARLOW SHAPLEY
1885–1972
American astronomer

JAN OORT
1900–92
Dutch astronomer

30-SECOND TEXT
Carolin Crawford

The solar system is contained in one of the spiral arms of our local galaxy, the Milky Way.

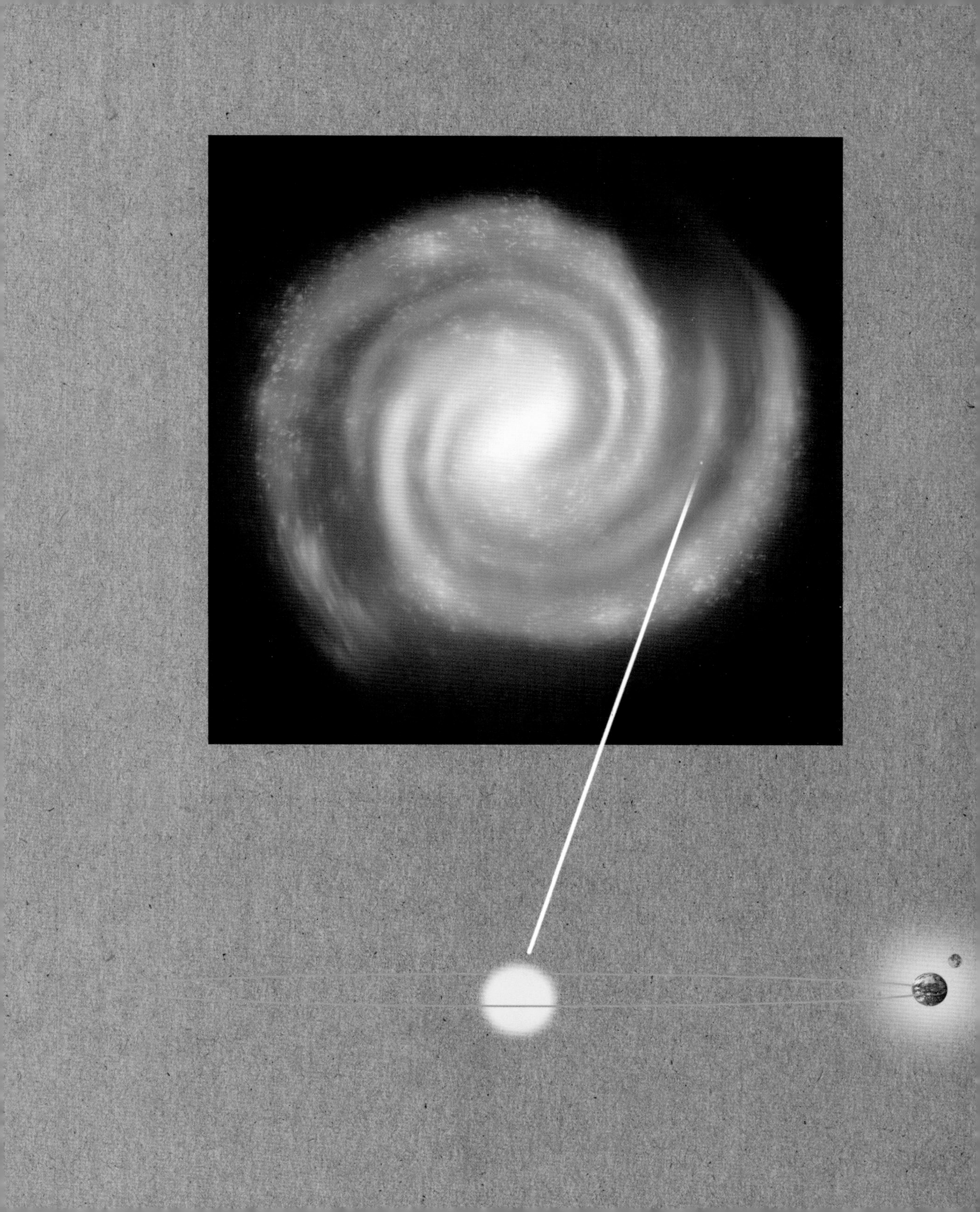

15 November 1738
Born in Hannover, Germany

1757
Emigrated to England

1766
Became organist at Octagon Chapel, Bath

1774
Began building telescopes and observing the skies, beginning with Orion Nebula

1780
Became director of the Bath orchestra

1781
Elected to the Royal Society

13 March 1781
Discovered what will become known as the planet Uranus

1782
Gave up music to become Court Astronomer

1782
Observed Messier objects; discovered the hitherto unseen Saturn Nebula

1783
Started regular sky sweeps

1783–1802
Observed and catalogued around 2,500 new nebulae and star clusters

1783
Published observations leading to the discovery of solar motion (theory that the solar system is moving through space)

1789
Finished his largest telescope, with a 1.2 m (48 in) aperture

1800
Discovered infrared radiation

1801
Met Napoleon Bonaparte and Charles Messier

1802
Published *Catalogue of 500 new Nebulae, nebulous Stars, planetary Nebulae, and Clusters of Stars; with Remarks on the Construction of the Heavens;* theorized that some double stars may be binaries orbiting each other

1803
Published *Account of the Changes that have happened, during the last Twenty-five Years, in the relative Situation of Double-stars; with an Investigation of the Cause to which they are owing*

1820
Cofounded Astronomical Society, which would receive its Royal Charter in 1831

25 August 1822
Died at Slough, Berkshire

WILLIAM HERSCHEL

The founder of modern stellar astronomy, discoverer of binary stars and the first person to realize that the solar system itself moves was not, amazingly, an astronomer by trade or training. Friedrich Wilhelm Herschel was born in Hannover to a family of musicians; he emigrated to England with his brother Jacob at the age of 19, and he spent four years playing and teaching the oboe, cello, harpsichord, violin and organ. In 1766 (his name now anglicized to William Herschel), he was appointed organist at the Octagon Chapel in Bath and settled into a musical career, composing 24 symphonies. It wasn't until he was 35 years old that he began to examine the skies. A musician's interest in English mathematician Robert Smith's *Harmonics* (1749) had led him to Smith's book *A Compleat System of Opticks* (1738), which piqued an interest in lenses and telescopes. Herschel made significant improvements to the Newtonian reflecting telescope of the day, quickly gaining an international reputation for his craftsmanship, building more than 400 models and developing a lucrative sideline in making and selling them. When he discovered Scottish astronomer James Ferguson's *Astronomy Explained on Sir Isaac Newton's Principles* (1756), and began looking through his telescopes to while away the long winter nights, the musician became the most significant astronomer of his age.

Noting his observations meticulously – with the help of his sister Caroline (herself no cosmic slouch; she found eight comets and at least four nebulae) – Herschel created an exhaustive catalogue of nebulae, multiple star clusters, single stars and deep-sky objects that is still in use today; by constantly revising, comparing and pruning his observations, he was, over a 25-year period, able to list more than 2,500 celestial objects, theorize (correctly) on the gravitational orbit of binary stars (of which he found 800) and work out that the solar system is moving, and in which direction (towards Lambda Herculis, a star in the constellation Hercules). On a March night in 1781, he discovered what would become known as the planet Uranus; he called it *Georgium Sidus* ('George's star') to honour the Hanoverian king of England, George III. His other achievements were many. He found two moons for Uranus in 1787 (later named Titania and Oberon), and two more moons for Saturn (Mimas and Enceladus), as well as demonstrating that the Milky Way must be disc-shaped. And while trying to find lenses suitable for studying the Sun, he discovered infrared radiation.

THE OTHER GALAXIES

the 30-second astronomy

3-SECOND BANG
The Milky Way is only one of about 100 billion galaxies scattered throughout the observable Universe – and the average galaxy contains about 100 billion stars.

3-MINUTE ORBIT
Measurement of a galaxy's distance from us relies on identification of an object within it for which we know the inherent brightness. Comparison of this expected luminosity to that observed yields the distance to the object's host galaxy. Such 'standard candle' objects include variable stars and supernovae. The furthest galaxies known are more than 13.2 billion light-years away and have a blobby appearance, suggesting they have assembled through the merging of smaller systems.

Although early telescopic observations of nebulae revealed that some displayed a curious spiral structure, it wasn't clear that they were not part of the Milky Way. Only in the early 1920s, when American astronomer Edwin Hubble determined the distance to the Andromeda nebula, was it proved that the spiral nebulae were separate, and, therefore, that the Milky Way was not the entirety of the Universe. There is a huge variety of galaxies, from dwarf to giant – ranging from one-thousandth to a thousand times the mass of the Milky Way, and from one-hundredth to ten times the Milky Way's size. The galaxies are generally classified according to their optical shape and contents. Spiral galaxies reveal bright lanes of active star formation that trace out a spiral pattern within the flattened disc surrounding a central bulge. The more common elliptical galaxies are ball-shaped structures rich in hot X-ray gas, but relatively devoid of the cold interstellar gas and dust required for massive star formation. Galaxies without any regular structure are known as 'irregular'; these are often the end-product of a gravitational interaction between two galaxies, and are what remains after a short-lived but dramatic 'starburst' phase.

RELATED TOPICS
See also
VARIABLE STARS
page 58

SUPERNOVAE
page 68

MOLECULAR CLOUDS & NEBULAE
page 78

3-SECOND BIOGRAPHIES
HEBER CURTIS & HARLOW SHAPLEY
1872–1942 & 1885–1972
American astronomers who publicly debated in 1920 about the size of the Universe

EDWIN HUBBLE
1889–1953
American astronomer

30-SECOND TEXT
Carolin Crawford

Galaxies vary in size and shape, from loose fluffy spirals to dense massive ellipticals. Ones that appear tiny (such as those arrowed) are usually the most remote.

GALACTIC STUCTURES

the 30-second astronomy

Many galaxies are tethered by their combined gravity to form clusters of galaxies. Hundreds, even thousands, of galaxies are packed into a volume of space a few tens of millions of light-years across. The first cluster to be noticed was included in the catalogue of nebulae prepared by French astronomer Charles Messier, which includes 11 'nebulae' towards the constellation of Virgo. Clusters were not systematically catalogued until the advent of detailed photographic plates in the 1950s, and were identified from overdensities of galaxies by eye. Most of the galaxies in a rich cluster are elliptical in shape, with a few blue spirals present at the outskirts of a cluster. The core is dominated by giant ellipticals, some of which are the most massive galaxies known. All the galaxies are bathed in a hot gaseous atmosphere that contains ten times more mass than that of the stars, but which is visible only at X-ray wavelengths. The observed physical properties of this gas, along with the motions of the galaxies in the cluster, and gravitationally lensed mirages of background sources, all point to most of the gravitating mass of a cluster being in the form of dark matter.

3-SECOND BANG
Galaxies are not distributed randomly across the sky, but collect together to form a tapestry of ever-larger structures.

3-MINUTE ORBIT
Many galaxy clusters are clumped into even larger structures called superclusters, which are usually flattened into slab-like concentrations known as walls. On a large scale, we see that the superclusters surround under-dense regions of comparable size, known as voids. Galaxy surveys show that this cobwebby pattern only repeats itself as we move outwards, giving the Universe a distinctly cellular appearance on the largest scales.

RELATED TOPICS
See also
THE OTHER GALAXIES
page 86

COSMIC X-RAYS
page 104

DARK MATTER
page 10

3-SECOND BIOGRAPHIES
HARLOW SHAPLEY
1885–1972
American astronomer

GEORGE O. ABELL
1927–83
American astronomer who catalogued clusters of galaxies

30-SECOND TEXT
Carolin Crawford

On the very largest scales of the Universe, galactic structures are distributed into long filaments that surround equally large voids in space. Massive clusters of galaxies are thought to form where the filaments intersect.

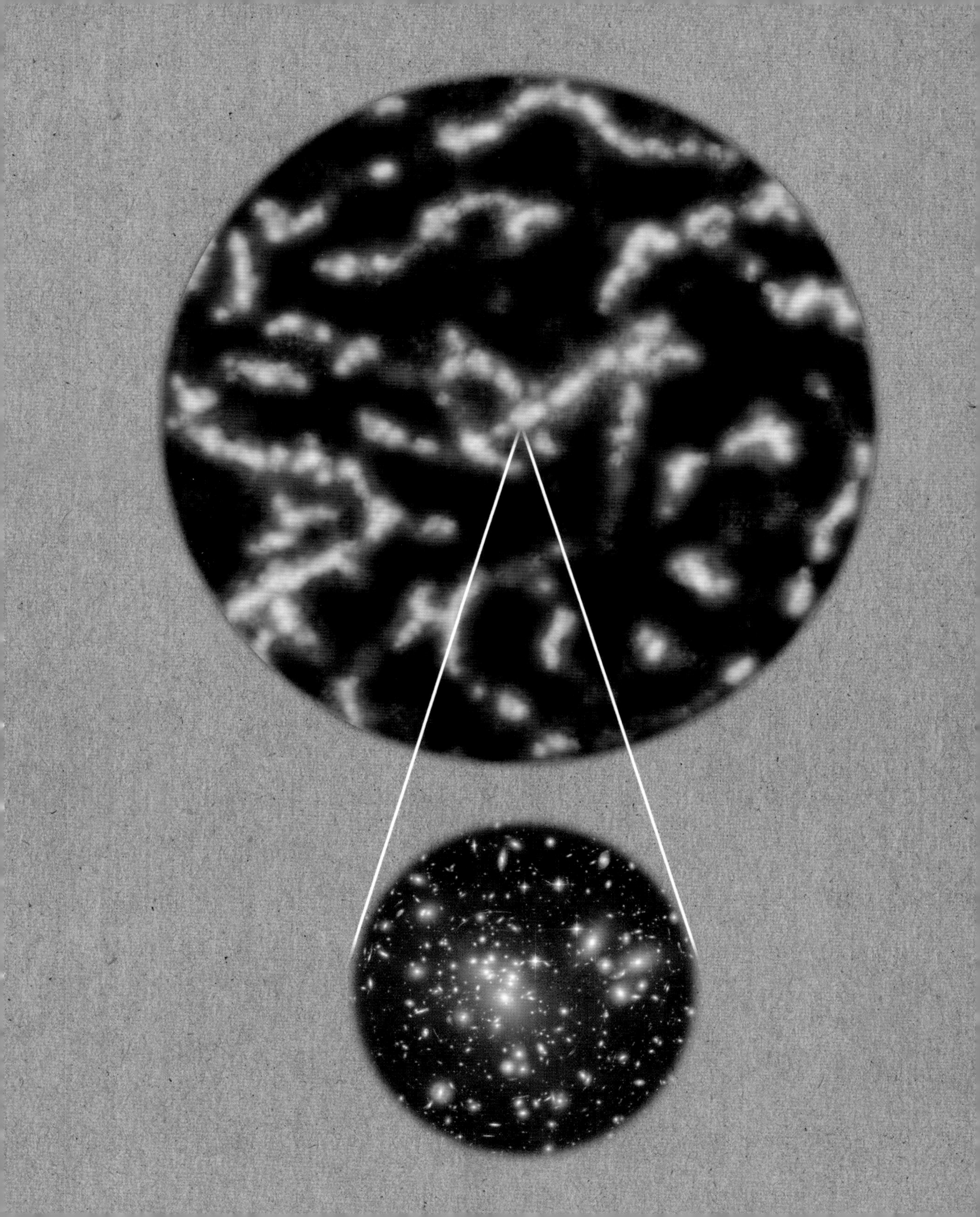

THE UNIVERSE

THE UNIVERSE
GLOSSARY

Big Bang The starting point of space and time in an explosion from a single, extremely hot and extremely dense point. According to various competing theories, the Universe whose expansion began with the Big Bang will end in the Big Chill, the Big Crunch or the Big Rip.

Big Chill Also known as 'the Big Freeze', one projected end for the expanding Universe, in which galaxies will drift away from one another, stars will burn out, galaxies will become exhausted, and the Universe itself will become vastly large, dark, and cold.

Big Crunch A second hypothesis for the end of the Universe, in which the Universe will expand until it reaches a critical point, then begin to contract, becoming more dense and hotter and finally reaching an infinitely dense, infinitely hot point called the Big Crunch. The Big Crunch could be the launching point for another Big Bang. The discovery of dark energy (a mysterious force fuelling the Universe's expansion) has made this scenario outmoded.

Big Rip A third hypothesis concerning the end of the Universe, in which matter (at all scales from galaxies to subatomic particles) will be torn apart by the dark energy that pushes the expansion of the universe.

Cosmic Microwave Background A diffuse field of radiation – the first light released in the aftermath of the Big Bang that spread out through the expanding universe. The discovery of the Cosmic Microwave Radiation in 1964 established the primacy of the Big Bang over other theories of the origin and state of the Universe.

cosmology The study of the birth, shape, growth, size and projected end of the Universe.

dark energy Energy that fuels the expansion of the Universe.

dark matter Unseen matter whose gravitational effects are detected on visible matter, galaxies, and large-scale structures in the Universe.

fundamental forces The four basic forces effective in the Universe: the gravitational force, the electromagnetic force, the strong nuclear force and the weak nuclear force.

Hubble Constant The rate of expansion of the Universe.

hypernova Extremely powerful explosion – releasing substantially more energy than a supernova – that gives issue to gamma ray bursts of long duration.

inflation Intensely short, very fast expansion that occurred in the aftermath of the Big Bang. It was followed by a period of relatively gradual expansion. Inflation is thought to have occurred for just an infinitesimally small fraction of a second, between 10^{-38} and 10^{-36} seconds after the Big Bang.

light-year The distance that light travels in one year: 9.5 trillion kilometres (5.9 trillion miles).

MOND Modified Newtonian dynamics, a theory suggesting that the effects of gravity may be longer-lasting and stronger than generally thought, and that these effects rather than dark matter hold together galaxies and other clusters that would otherwise fall apart.

quasars Quasi-stellar radio sources. Astronomers first thought these were radio stars, but then discovered that the radio waves were emitted by a galaxy with a bright nucleus consisting of a supermassive black hole.

radio stars Stars that emit radio waves, such as pulsars.

steady-state theory Proposal first put forward by British physicist James Jeans in around 1920 and developed in 1948 by British astronomer Fred Hoyle and colleagues to counter the rival Big Bang theory of the origins and state of the Universe. According to the steady-state theory, new matter is being continuously created in a constantly expanding Universe, forming new stars and galaxies, while old galaxies and stars become unobservable as the Universe expands. The steady-state universe has a constant average density and no end or beginning in time. This theory has been discredited.

THE BIG BANG

the 30-second astronomy

An obvious consequence of the discovery that space is expanding is that there must have been a beginning to the Universe. All matter, space and time came into being at a unique starting point that we call the Big Bang. This idea was proposed initially by Georges Lemaître as a possible solution to Albert Einstein's equations of General Relativity; it gained universal acceptance with the discovery of the Cosmic Microwave Background in 1964. Subsequent observations of the way that the population of galaxies containing strong radio sources changed with time further supported the idea of an evolving Universe. Astronomers have no firm explanation of what triggered the Big Bang, as our current understanding of the laws of physics cannot describe such an extremely hot and dense phase of matter – let alone yield any description of what could have happened 'before' this event. Within a tiny fraction of a second, the Universe underwent a short period of inflation, resulting in an extraordinarily rapid increase in its size – leading to its subsequent continued expansion – and its contents cooling. The first elementary particles began to form, and the fundamental forces separated out to take on their current nature.

3-SECOND BANG

Everything in our Universe is thought to have originated in an event we term the 'Big Bang', which marks the start of space and time.

3-MINUTE ORBIT

Ironically, the term 'Big Bang' was first coined as a disparaging description by one of its most vocal opponents, Fred Hoyle. Although he promoted the alternative 'steady-state' cosmology of a continuous creation of matter, Hoyle – and his colleagues – showed that the large and uniform amount of helium everywhere in the cosmos must have been created in a primordial Universe instead of solely through nuclear reactions in stars.

RELATED TOPICS

See also
THE EXPANDING UNIVERSE
page 96

COSMIC MICROWAVE BACKGROUND
page 100

3-SECOND BIOGRAPHIES

ALEXANDER FRIEDMANN
1888–1925
Russian/Soviet mathematician and physicist

GEORGES LEMAÎTRE
1894–1966
Belgian astronomer

FRED HOYLE
1915–2001
British astronomer

MARTIN RYLE
1918–84
British radio astronomer

30-SECOND TEXT

Andy Fabian

The Big Bang is a term used to describe the monumental event 13.7 billion years ago that triggered the creation of the Universe.

THE EXPANDING UNIVERSE

the 30-second astronomy

Astronomer Edwin Hubble's second major breakthrough, after he established that there were galaxies beyond the Milky Way, was the discovery that the Universe is expanding. Nearly all galaxies are moving away from us, with more distant galaxies receding faster. This is the signature of an expanding Universe: the space separating galactic structures stretches and pushes them away from each other. By tracking this expansion back through time, astronomers deduced a finite age for the Universe of 13.7 billion years. Recent observations of supernova explosions in remote galaxies, the size of structures within the Cosmic Microwave Background, and the matter content of galaxy clusters have all shown that the rate of expansion has increased over the last 6 billion years. This accelerated expansion requires there to be more to the Universe than the radiation, ordinary matter and dark matter that we observe. The missing ingredient, which is thought to provide the accelerating push, is referred to as 'dark energy', and may account for three-quarters of the total contents of the cosmos. Determining its exact nature will inform us whether the eventual fate of the Universe in many billions of years' time is a 'Big Chill' or a 'Big Rip'.

3-SECOND BANG

Tracking the motions of galaxies reveals that our Universe is not static and eternal, but is evolving and growing ever more rapidly with time.

3-MINUTE ORBIT

The rate of expansion of the Universe, known as the 'Hubble Constant', was uncertain (by a factor of two) until the mid-1990s, when observations of individual stars in nearby galaxies made with the Hubble Space Telescope enabled astronomers to deduce the Hubble Constant with a precision of a few per cent. It is the expansion of the Universe that is responsible for the light from distant objects appearing redshifted.

RELATED TOPICS

See also
THE OTHER GALAXIES
page 86

THE BIG BANG
page 94

3-SECOND BIOGRAPHIES

SAUL PERLMUTTER
1959–
American astrophysicist. He, Schmidt and Riess showed that the expansion of the Universe is speeding up

BRIAN SCHMIDT
1967–
American-Australian astrophysicist

ADAM RIESS
1969–
American astrophysicist

30-SECOND TEXT

Andy Fabian

Since the Big Bang, space has been expanding, stretching the distances between galactic structures, and at a rate that has been increasing over the last 6 billion years.

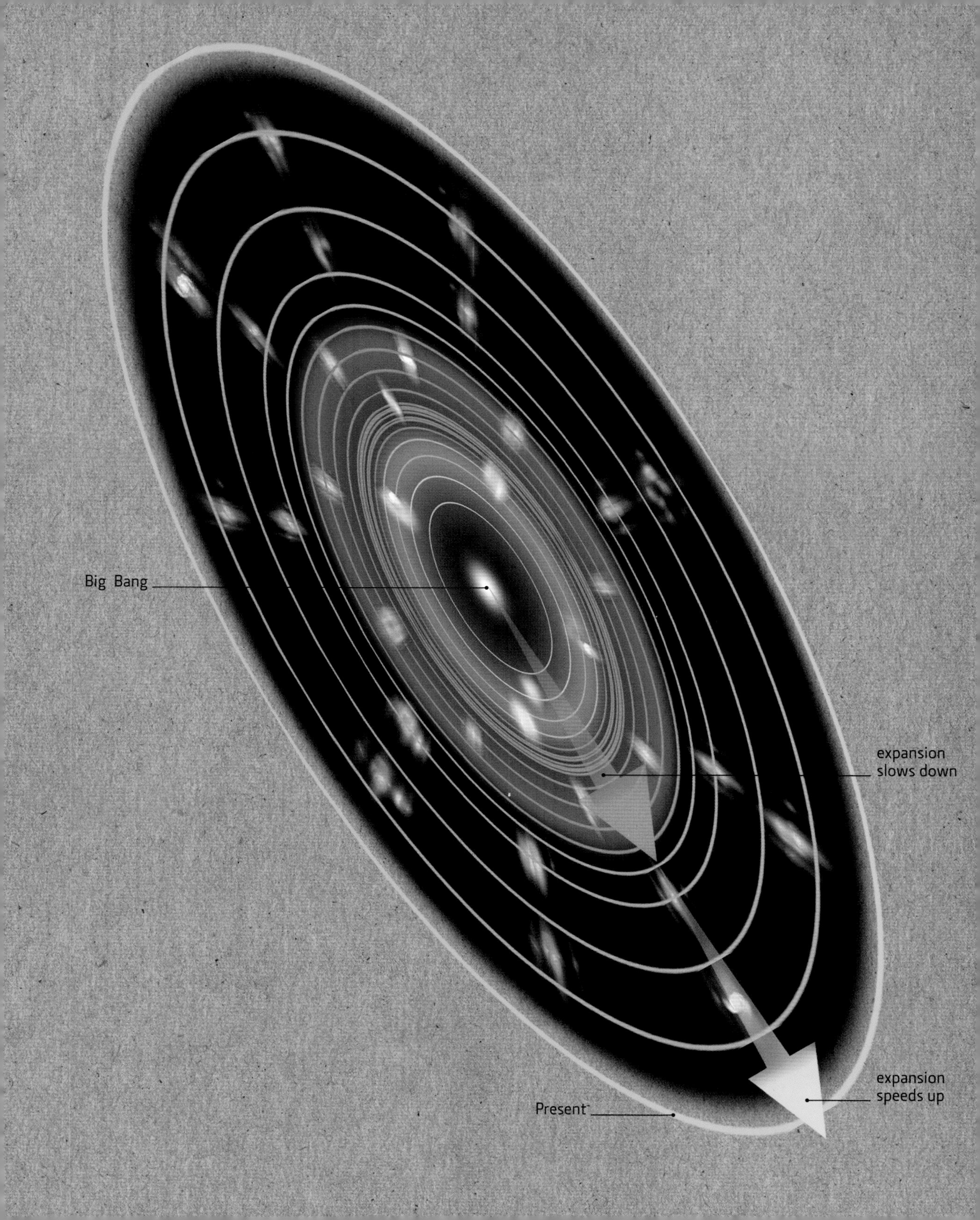

Big Bang
expansion slows down
expansion speeds up
Present

20 November 1889
Born in Mansfield, Missouri

1898
Family moved to Chicago

1906–10
Studied mathematics, astronomy, and science at the University of Chicago

1910–13
Rhodes Scholar at Queen's College, Oxford, studying law, literature and Spanish

1913
Returned to the USA; brief career as a lawyer in Louisville, Kentucky

1914–17
Studied for PhD at the University of Chicago; dissertation entitled 'Photographic Investigations of Faint Nebulae'

1917
Offered post at Mount Wilson Observatory, Pasadena, California, but turned it down to enlist and fight in the First World War

1917–18
Served in US army, rising to rank of major

1919
Took up post at Mount Wilson Observatory

1923
Discovered Cepheid variable in Andromeda nebula (M31)

1926
Devised a method of classifying galaxies (the Hubble sequence)

1929
Formulated Redshift Distance Law of Galaxies (Hubble's Law)

1935
Discovered asteroid 1373 Cincinnati; wrote *The Observational Approach to Cosmology and The Realm of the Nebulae*

1940
Awarded the Gold Medal of the Royal Astronomical Society

1942–45
Served in US army; based in Aberdeen, Maryland

1946
Awarded Medal of Merit for work on ballistics

1948
Made Honorary Fellow of Queen's College, Oxford

1949
First person to use the Hale telescope (the world's largest-aperture optical telescope at the time) at Mount Palomar, San Diego, California

1949
Suffered a heart attack

28 September 1953
Died in San Marino, California

1990
NASA launched the Hubble Space Telescope, named in his honour

EDWIN HUBBLE

Edwin Powell Hubble was an all-American boy from the Midwest – smart, clever, ambitious, strong and an all-round sportsman, with exactly the mixture of brain and brawn that is likely to win you a Rhodes Scholarship to study overseas at the University of Oxford in England. (Sure enough, Hubble won a scholarship and spent three years at Queen's College, Oxford.) Back home, he was adored at the Indiana high school at which he taught Spanish, mathematics, physics and basketball for a year, tried a brief spell as a lawyer (to please his father), and proved himself more than willing to do his patriotic duty in both world wars. Underneath it all, however, he was a stargazer. Astronomy was his first love, and when he gave up his legal practice to go back to Chicago University to study for a PhD , he said: 'I knew that even if I were second or third rate, it was astronomy that mattered.'

However, he was neither a second- nor third-rate astronomer. Hubble's discoveries opened up the universe. He led what Professor Stephen Hawking has called 'one of the great intellectual revolutions of the 20th century'. Working with the telescopes at Mounts Wilson and Palomar, California, Hubble proved that we are surrounded by millions of galaxies. (Herschel had suspected it, but Hubble proved it.) Hubble introduced a method of classifying galaxies, based on their apparent shape (elliptical, lenticular, spiral and irregular), now known as the Hubble sequence. By careful measurements of galactic redshift (the way the wavelength of light from a galaxy lengthens towards the red end of the colour spectrum because the galaxy is moving away), he demonstrated that galaxies are moving away from each other at a constant rate (Hubble's constant). He called this the Redshift Distance Law of Galaxies, but it is now known as Hubble's Law. Furthermore, he showed that if galaxies are moving apart, then the Universe must be expanding. This discovery backed up the Belgian astronomer Georges Lemaître's Big Bang theory, introduced the previous year, and proved so impressive that Albert Einstein came to visit Hubble in 1931, specifically to congratulate him. Hubble had transformed astronomy into cosmology, and became known as the 'pioneer of distant stars'.

COSMIC MICROWAVE BACKGROUND

the 30-second astronomy

The very early Universe was an incredibly hot soup of charged particles and photons of light. These were so densely packed that any individual photon could not travel far without undergoing some kind of interaction, and so no light could escape. As it expanded, the Universe cooled down, until – around 380,000 years after the Big Bang – it reached temperatures low enough that the charged particles combined to form the very first atoms. These no longer interfered with the passage of the photons, which became free to stream out across the Universe. They became a diffuse cosmic background radiation, which uniformly fills the whole sky. The intervening expansion of space has caused the wavelength of the (originally energetic) photons to be stretched by enormous factors. The radiation is observed now to glow brightest in microwave light, corresponding to a temperature of only 2.725 degrees above absolute zero. Although its existence had been theorized since the 1940s, the serendipitous discovery of the background radiation in 1964 by Arno Penzias and Robert Wilson provided the clinching piece of evidence in support of the Big Bang.

3-SECOND BANG
The Cosmic Microwave Background – a snapshot of the earliest light released after the Big Bang – informs our ideas about the origin and structure of the Universe.

3-MINUTE ORBIT
Precision satellite measurements of the cosmic microwave background radiation map its temperature instead of its intensity, and reveal a mottled appearance due to minute variations away from the average. Such fluctuations trace slight overdensities in the energy/matter mix, which can provide a focus for gravitational attraction: these 'seeds' are thought to develop, much later, into giant structures, such as the galaxies that occupy the present-day Universe.

RELATED TOPICS
See also
THE BIG BANG
page 94

THE EXPANDING UNIVERSE
page 96

BEYOND VISIBLE LIGHT
page 102

3-SECOND BIOGRAPHIES
ARNO PENZIAS
1933–
American physicist

ROBERT WILSON
1936–
American physicist

30-SECOND TEXT
Andy Fabian

The CMB is a glimpse of the earliest observable Universe, at the point where matter first began to condense under gravity to sow the seeds of galaxy formation.

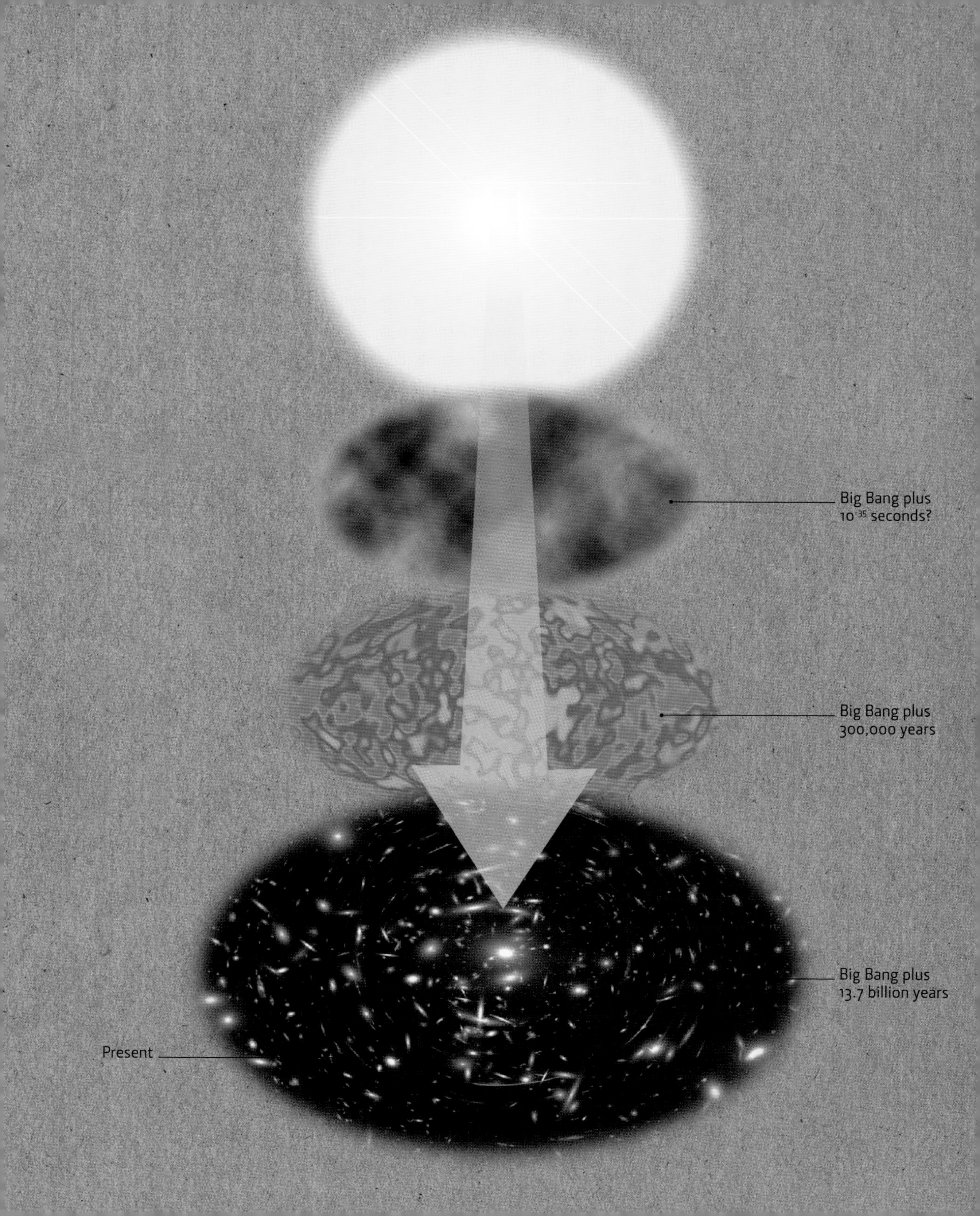

Big Bang plus
10^{-35} seconds?

Big Bang plus
300,000 years

Big Bang plus
13.7 billion years

Present

BEYOND VISIBLE LIGHT

the 30-second astronomy

3-SECOND BANG
We can only see a tiny fraction of the light around us, but to fully understand the universe we need to see beyond visible light.

3-MINUTE ORBIT
Only light of certain energy (or wavelength) can pass through our atmosphere. Radio waves and visible light can pass straight through, allowing us to admire the night sky with our own eyes or with a radio dish. But harmful X-rays and gamma rays are blocked by the atmosphere, and so we need telescopes in space to see what violent phenomena can create these most energetic forms of light.

Humans see light between the colours of red and blue, but our eyes are not sensitive to anything beyond these colours. We cannot see infrared light beyond red, nor can we see ultraviolet, X-ray or gamma-ray light beyond blue. This is a hindrance to astronomers – all objects in the universe emit light, but not necessarily the visible light that we can see; the wavelength depends on temperature. When stars light up for the first time, before nuclear fusion has begun to warm them up beyond a thousand degrees or so, they emit infrared light. Once fusion starts and the stars warm up to temperatures of a few thousand degrees, they are hot enough to emit visible light, like an incandescent lightbulb. Heat up the gas to a few hundred thousand degrees, and it will emit ultraviolet light. Gas at a few million degrees emits X-ray light; this is the tell-tale sign that violent processes are afoot. And when massive stars collapse to form black holes, the gas is heated up to a few billion degrees, so hot that it emits gamma-ray light.

RELATED TOPICS
See also
COSMIC MICROWAVE BACKGROUND
page 100

COSMIC X-RAYS
page 104

THE LIGHT SPECTRUM
page 122

3-SECOND BIOGRAPHIES
ISAAC NEWTON
1642–1727
English physicist

WILLIAM HERSCHEL
1738–1822
German-born British astronomer

30-SECOND TEXT
Darren Baskill

Because gas at different temperatures emits light at different wavelengths, to fully understand our universe we need to look for light at all wavelengths.

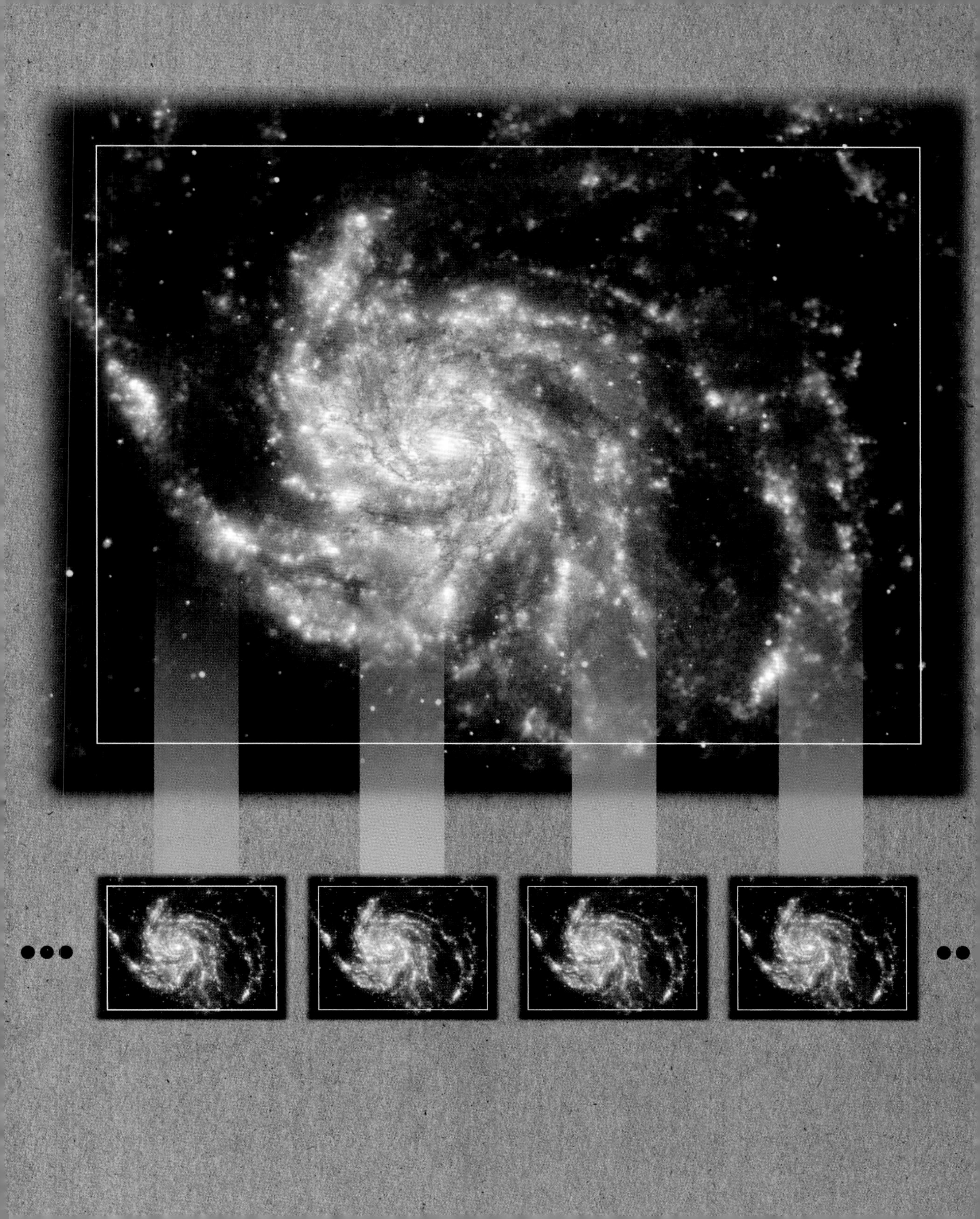

COSMIC X-RAYS

the 30-second astronomy

On 18 June 1962, a rocket flight above New Mexico detected X-rays coming from beyond the solar system, but the sources of the X-rays were unknown at the time. X-rays are a type of high-energy light, and are emitted from gas heated up to more than 1 million degrees through violent processes. Fifty years on, we have a far more complete understanding of where cosmic X-rays originate, as the latest space telescopes reveal the X-ray Universe in detail. We now know that the X-rays detected in 1962 were created by gas being torn from a star less than half the mass of the Sun crashing down onto a dense neutron star. Cosmic X-rays reveal how violent our universe is. Dead stars in the heart of the remains of supernovae, gas plummeting into small black holes within our galaxy, and supermassive black holes at the heart of other galaxies all reveal themselves by emitting X-rays. Other X-ray sources include stars being torn apart by their partners; the most compact and dense stars known, white dwarf and neutron stars; and million-degree hot gas in the largest structures in the Universe, massive galaxy clusters.

3-SECOND BANG
In the most extreme parts of our Universe, gas heated to millions of degrees emits X-ray radiation, giving us a glimpse of violent events.

3-MINUTE ORBIT
Our atmosphere blocks out harmful X-rays, so to see them we need to go into space. X-ray telescopes are sometimes launched on brief five-minute rocket flights, but while excellent for testing technology, these only provide a glimpse of what is happening. For a deeper understanding, Earth-orbiting observatories – such as NASA's Chandra, Europe's XMM-Newton, and the Japanese Suzaku observatories – are used to monitor the violent phenomena that emit X-rays.

RELATED TOPICS
See also
SUPERNOVAE
page 68

BLACK HOLES
page 70

BEYOND VISIBLE LIGHT
page 102

3-SECOND BIOGRAPHIES
RICCARDO GIACCONI
1931–
Italian-American astrophysicist

BRUNO ROSSI
1905–93
Italian-American pioneer of X-ray astronomy

30-SECOND TEXT
Darren Baskill

At the centre of this galaxy, an explosion with the energy of several hundred thousand supernovae has heated gas to such high temperatures that it emits X-rays.

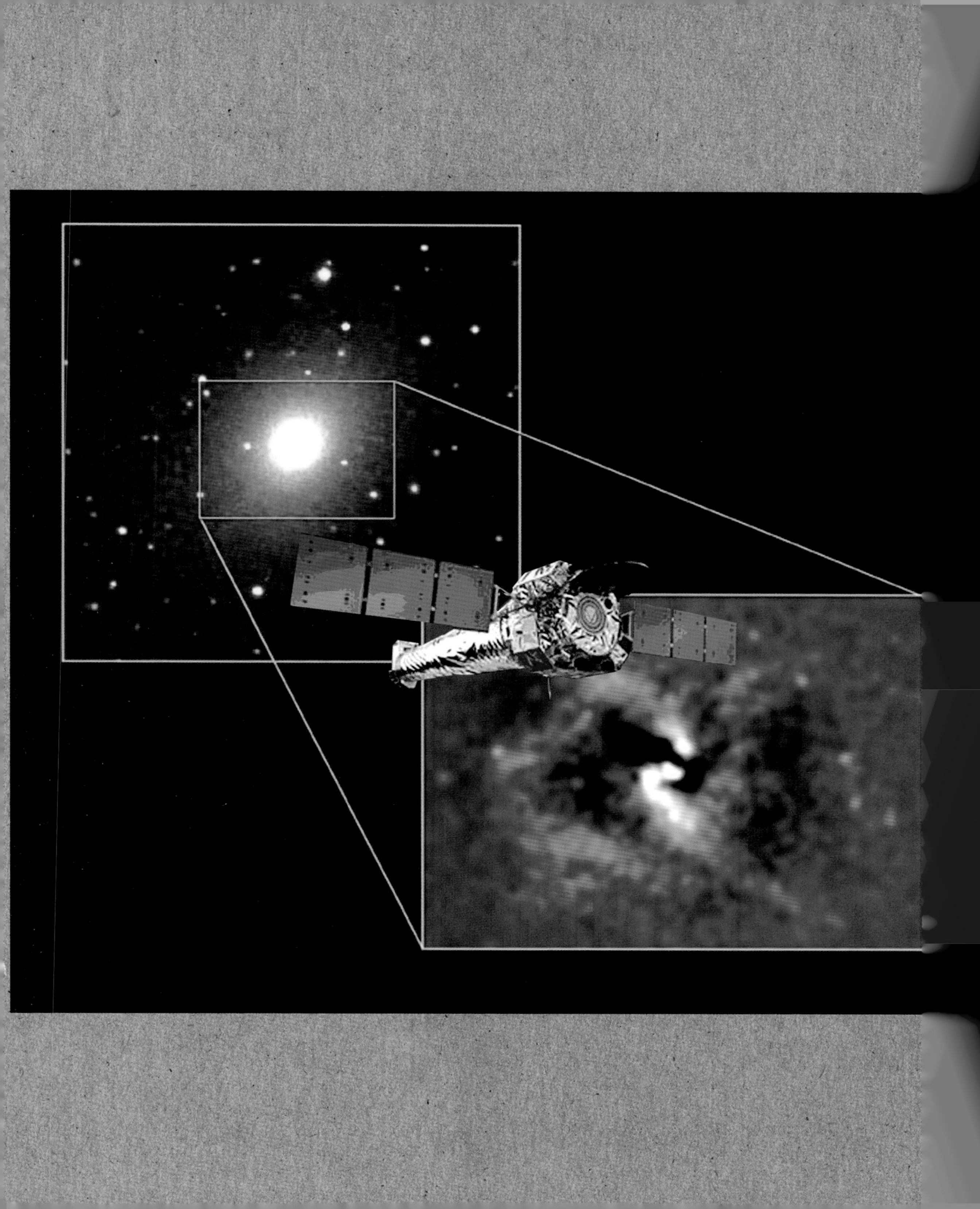

GAMMA RAY BURSTS

the 30-second astronomy

3-SECOND BANG
Gamma ray bursts are celestial explosions – the biggest bangs in the Universe since the Big Bang, happening somewhere in the Universe at least daily.

3-MINUTE ORBIT
The longer type of gamma ray burst is from the explosion of a hypernova, but the origin of the shorter bursts remains mysterious. Some astronomers hypothesize that the shorter bursts are generated when a neutron star is swallowed by a black hole, or when two neutron stars collide and merge to make a black hole.

Gamma ray detectors on board orbiting space satellites detect, about once per day, a brief, intense burst of gamma rays coming from a celestial source. Each burst comes from a distant galaxy, and causes an afterglow of X-rays and light, and a longer-lasting radio afterglow. The bursts are enormously powerful explosions, ten times the energy of a typical supernova, which is why astronomers can detect them in galaxies on the far side of the universe. There are two types of gamma ray burst: one lasting less than a second, the other perhaps half a minute. In some instances of the longer gamma ray burst, a supernova has been seen to appear afterwards, in the same galaxy. The afterglow is caused by collisions of the material thrown out by the supernova into its surrounding gas. The gamma rays are emitted when a very massive, quickly rotating star collapses to form a black hole. Rotation clears out the space along the axis, and the gamma rays escape in a beam. We see the beam only if it is pointed to Earth, so 'daily' is an underestimate of how often these events – known as 'hypernovae' – actually occur.

RELATED TOPICS
See also
PULSARS
page 64

SUPERNOVAE
page 68

BLACK HOLES
page 70

3-SECOND BIOGRAPHY
RAY KLEBESADEL
1932–
American defence scientist who serendipitously discovered gamma ray bursts

30-SECOND TEXT
Paul Murdin

Gamma ray bursters convert their explosive force into energetic radiation of all kinds, and beam gamma rays across the Universe.

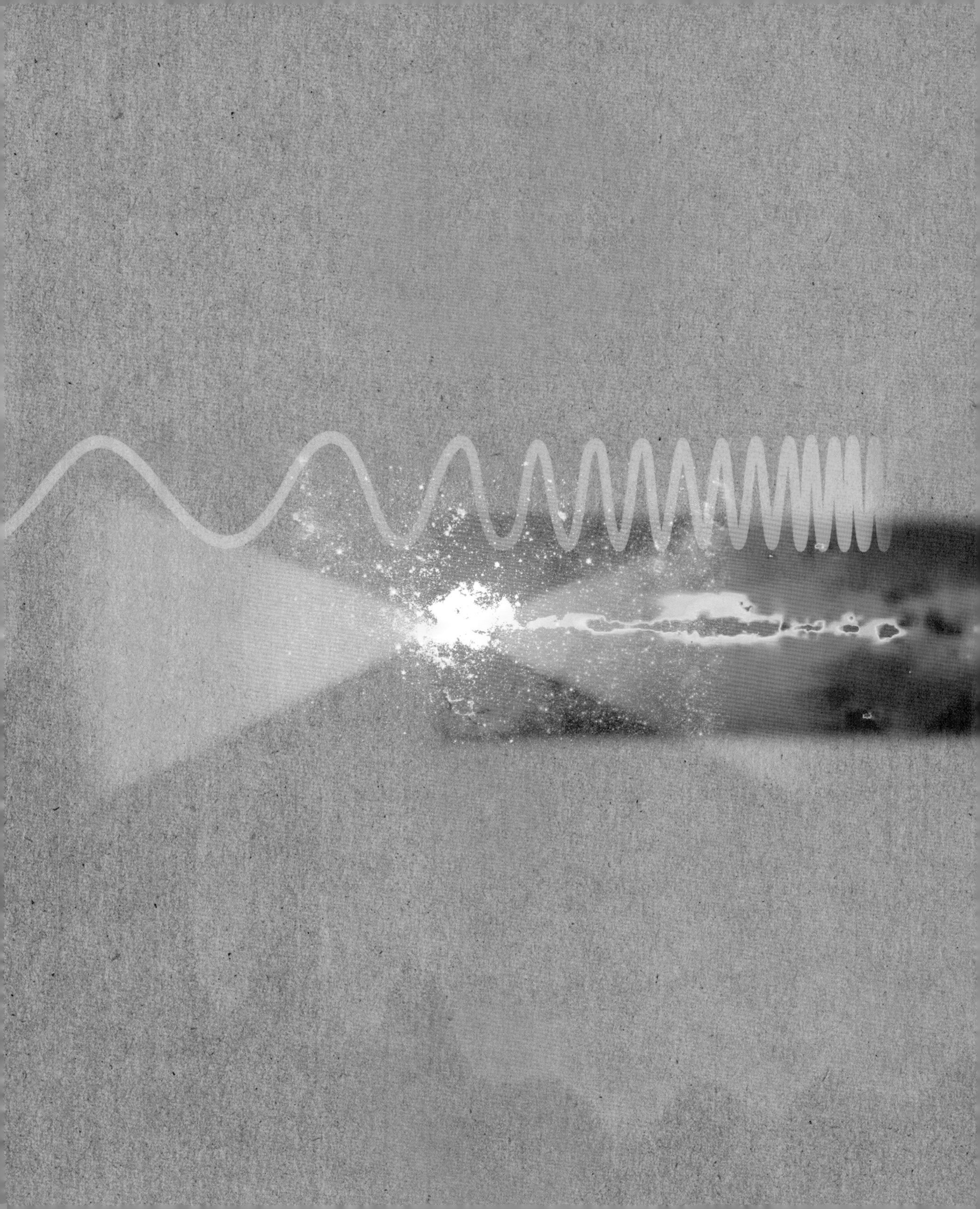

QUASARS

the 30-second astronomy

Quasars were identified when astronomers discovered radio sources that coincided with what looked like ordinary stars but turned out to be galaxies with bright nuclei. The phrase used to describe these phenomena, 'quasi-stellar radio sources', was subsequently abbreviated to 'quasars'. Images of exquisite precision made by radio telescopes, together with other lines of evidence, indicate that the nuclei of quasars are very small – the size of our solar system. Gas and dust circulate around the nucleus at very high speeds, orbiting something that, although small, is massive – perhaps millions or billions of times the mass of the Sun. The nucleus of a quasar is a supermassive black hole; the source of the quasar's power is the release of energy as gas dribbles into the black hole. Sometimes an individual star may fall in as a single episode. The tidal force of the black hole draws the star out into a gaseous filament, and a great deal of energy blazes out in a bright flare. So much energy is released by some supermassive black holes that some of the infalling gas is ejected. The gas speeds out in antiparallel jets, aligned in opposite directions, extending far into intergalactic space.

3-SECOND BANG
A quasar is a galaxy with a bright nucleus that consists of a super-massive black hole, which radiates typically the power of 1,000 ordinary galaxies.

3-MINUTE ORBIT
Most galaxies probably have a central black hole, although most are not quasars. The infall of gas that makes a galaxy a quasar can be triggered by the chance close approach of another galaxy. The black hole in our own galaxy, 4 million times the Sun's mass, is dormant. When the Andromeda Galaxy passes near to ours in a few billion years, the sleeping giant will probably awake; our galaxy will become a quasar.

RELATED TOPICS
See also
BLACK HOLES
page 70

3-SECOND BIOGRAPHIES
MAARTEN SCHMIDT
1929–
Dutch-American astronomer who first identified a quasar

30-SECOND TEXT
Paul Murdin

Gas swirls around a black hole in the centre of a galaxy. The gas that falls in emits an amazing amount of power.

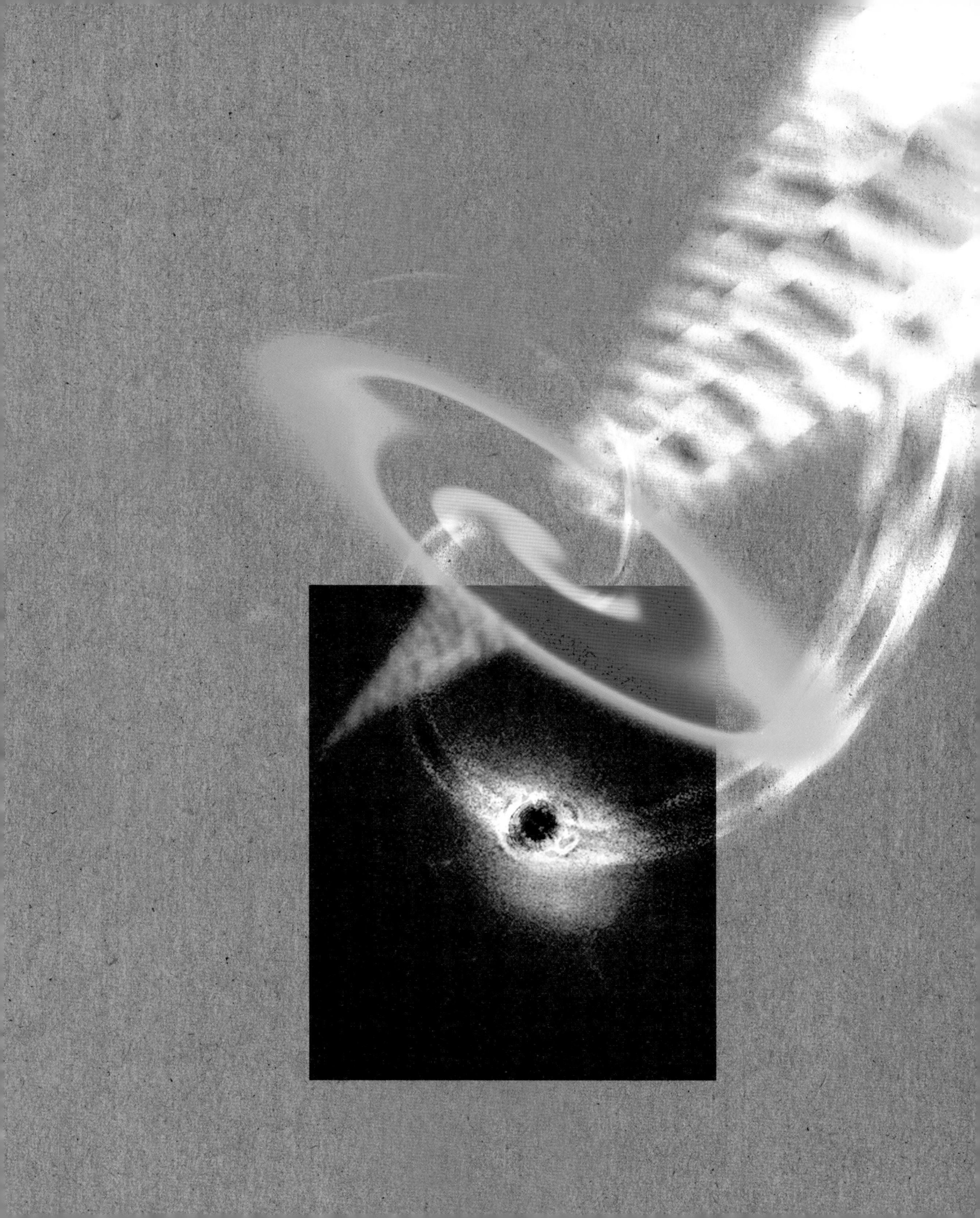

DARK MATTER

the 30-second astronomy

3-SECOND BANG
Astronomers know that dark matter exists because they can observe its gravitational influence, but they cannot see dark matter itself or define what it is.

3-MINUTE ORBIT
Scientists trying to explain dark matter have produced some intriguing theories: MACHOs (MAssive Compact Halo Objects), emitting no radiation and very difficult to identify, could exert a gravitational pull at the outskirts of galaxies; another suggestion is that WIMPs (Weakly Interacting Massive Particles) are responsible for exerting this invisible gravitational pull. WIMPs – individually tiny, but massive in large quantities – are the favoured explanation, but physicists have yet to prove they exist.

Our Sun orbits the Milky Way Galaxy, the massive city of stars in which we live, once every 250 million years. Stars close to the black holes at the heart of galaxies orbit quickly and this prevents them from falling in; further out, we would expect to find that stars either move at a more sedate pace or fly off out into deep space – but they move quickly without leaving their orbit. One explanation for this discrepancy is that there must be large amounts of matter, with a powerful gravitational pull, holding onto these stars. Astronomers call this invisible matter 'dark matter': according to the theory, dark matter makes up 83 per cent of the matter in the known Universe. Dark matter also reveals itself on even larger scales, when we look at how galaxies orbit each other in clusters. Proving what dark matter is constitutes one of the biggest challenges of twenty-first-century astrophysics. Experiments have been set up to try to detect hints of elusive particles that may be the source of dark matter; most of these experiments are placed deep underground to shield them from cosmic rays that would otherwise saturate the sensitive detectors.

RELATED TOPICS
See also
THE BIG BANG
page 94

THE EXPANDING UNIVERSE
page 96

DARK ENERGY
page 112

3-SECOND BIOGRAPHIES
FRITZ ZWICKY
1898–1974
Swiss astronomer

JAN OORT
1900–92
Dutch astronomer

30-SECOND TEXT
Darren Baskill

Dark matter is a source of gravity that is holding our galaxy – and our universe – together. But as yet, no one knows what dark matter is.

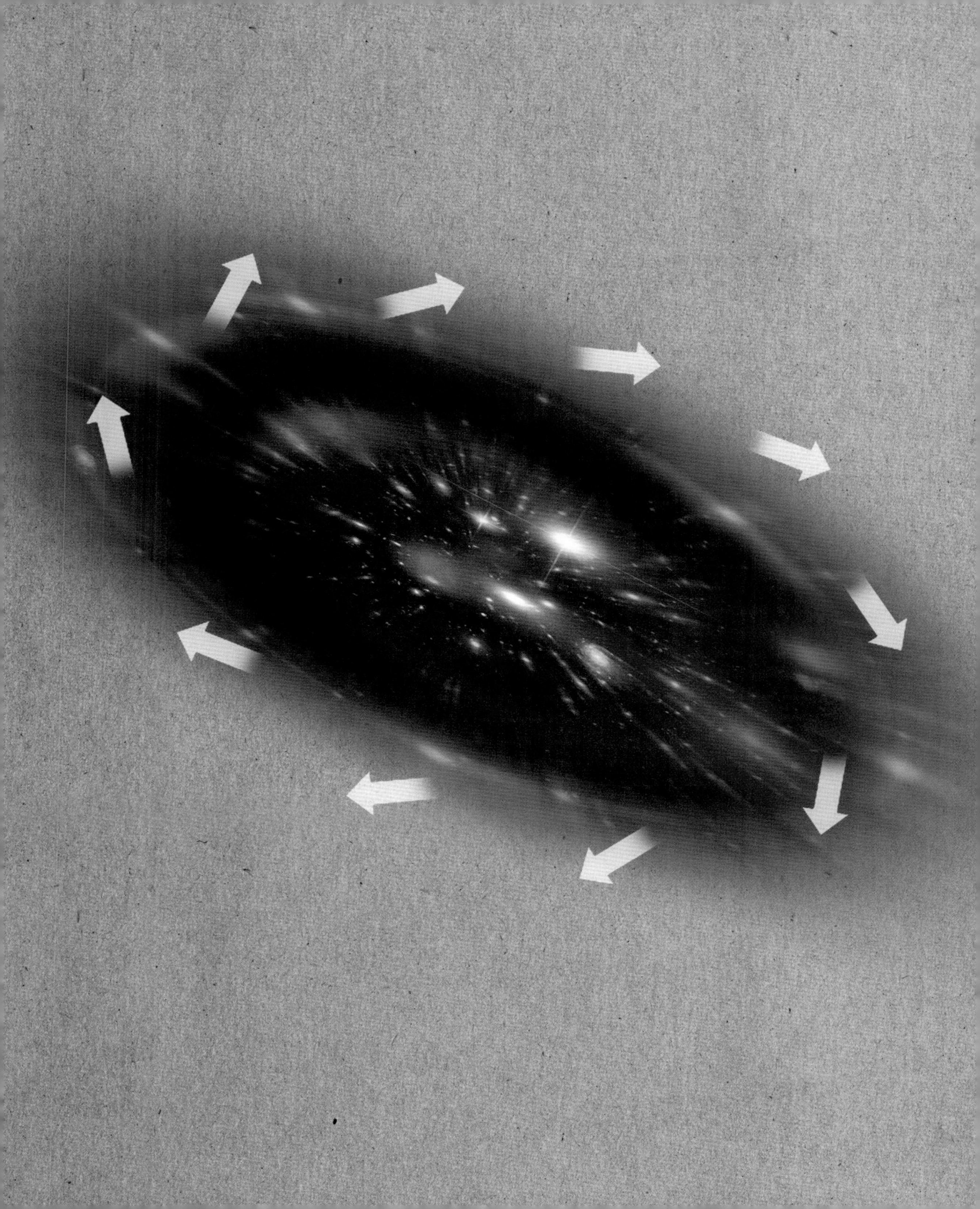

DARK ENERGY

the 30-second astronomy

Early in the twentieth century – before astronomers discovered that the Universe is expanding – Albert Einstein attempted to describe a universe of static galaxies, in his General Theory of Relativity. However, there was a problem: galaxies have a mutual gravitational attraction, and this should mean that a static universe would collapse; to explain why it did not, Einstein arbitrarily invented an outward push using a concept that he called the cosmological constant, Λ. When it was discovered that the universe was expanding, Einstein abandoned this idea, describing it as 'my greatest blunder'. Galaxies do attract each other, slowing down the expansion of the Universe, so distant galaxies, viewed as they were long ago, should be expanding faster than nearby ones. In 1998–99, it became possible with the Hubble space telescope to measure objects, such as Type 1a (a kind of supernova), so distant that their light has taken billions of years to reach us. To everybody's surprise, the distant galaxies inhabited by Type 1a supernovae expand more slowly than galaxies do now. The expansion of the Universe is actually accelerating: space is releasing 'dark energy' to push the Universe to expand faster. Einstein had thought of something important after all.

3-SECOND BANG

The Universe is expanding, with galaxies separating at higher and higher speeds – 'dark energy' is energy released from space to push this expansion of galaxies.

3-MINUTE ORBIT

Dark energy is one of a number of discoveries that illustrate that space is not just a passive volume of nothing, but instead an active physical entity, creating pairs of particles, bending and distorting light, sustaining waves of attraction that propagate from one mass to another and, indeed, creating a universe in a Big Bang. Space does interesting things, just like matter.

RELATED TOPICS

See also
SUPERNOVAE
page 68

THE EXPANDING UNIVERSE
page 96

3-SECOND BIOGRAPHIES

SAUL PERLMUTTER
1959–
American astrophysicist; codiscoverer, with Schmidt and Riess, of the acceleration of galaxies and the phenomenon of dark energy

BRIAN SCHMIDT
1967–
Australian astrophysicist

ADAM RIESS
1969–
American astrophysicist

30-SECOND TEXT

Paul Murdin

Our universe's history reads left to right as the Big Bang, the cooling afterglow, the formation of galaxies, and their expansion, which gets faster through the release of dark energy.

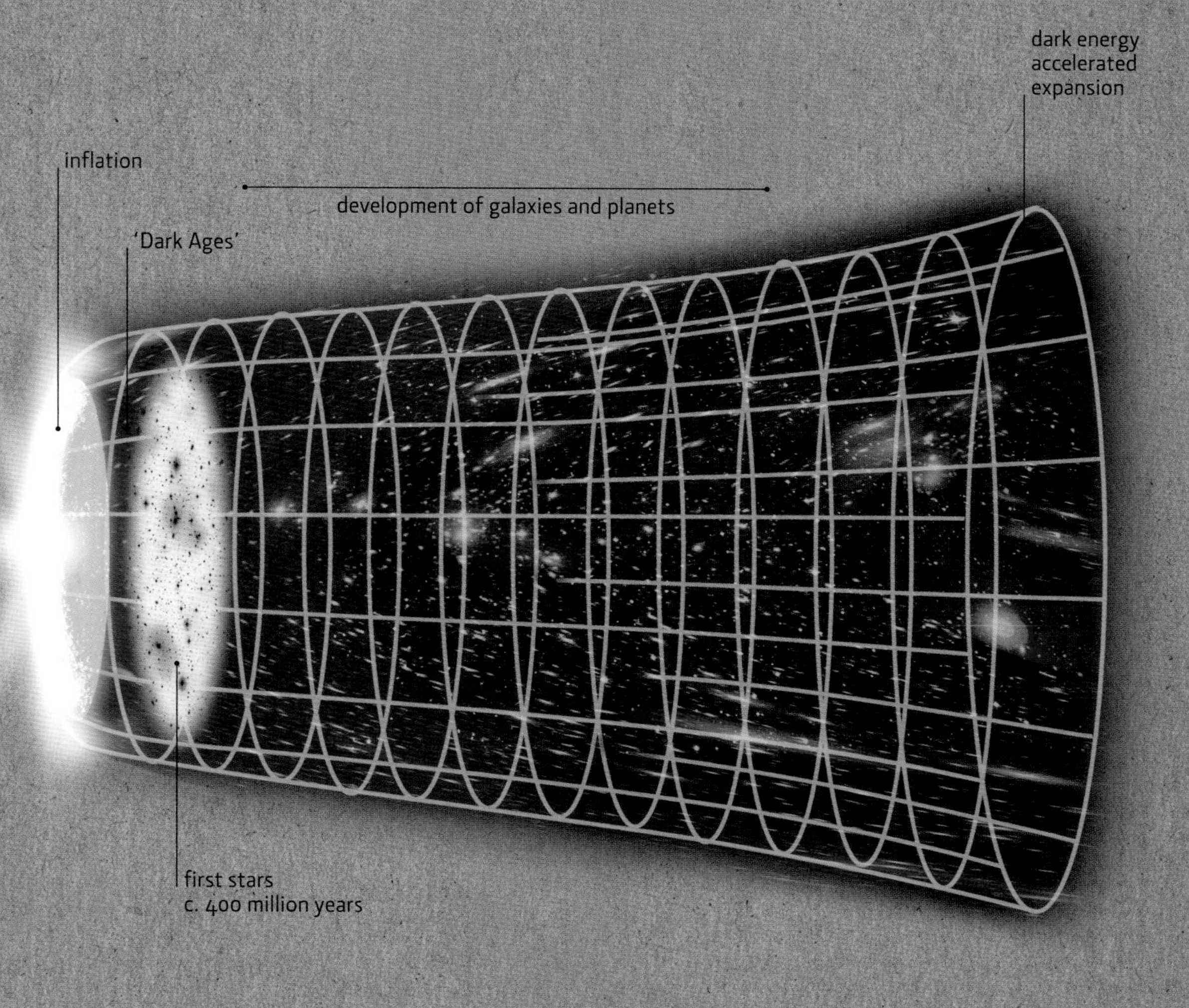
dark energy
accelerated
expansion
inflation
development of galaxies and planets
'Dark Ages'
first stars
c. 400 million years
13.7 billion years

SPACE & TIME

SPACE & TIME
GLOSSARY

61 Cygni Binary star in the constellation Cygnus. It was the first star to have its parallax measured, in 1838, by Prussian astronomer Friedrich Bessel, and when Bessel calculated its distance at 10.4 light-years, this was the first estimate of distance to a star other than our Sun. The actual distance is 11.4 light-years.

accretion The capturing and drawing in of gas by a massive body. As the gas is captured and spirals inwards towards a black hole, temperatures of millions of degrees are generated and the gas gives off X-ray radiation. Astronomers are able to identify the presence of black holes, which are invisible, from this radiation. Accretion also describes the capturing of gas or other matter by a small star (or remains of a star) from its partner in a binary star system.

asteroid A rocky body in our solar system that is smaller than a planet and in orbit around the Sun. The majority of identified asteroids are found in the asteroid belt between the orbits of Mars and Jupiter. Research in 2012, led by the Carnegie Institution in Washington DC, suggested that landings on Earth of asteroids – and not comets, as previously thought – were the original source of Earth's water.

blueshift Compression of light's wavelength towards the blue end of the spectrum, caused by the fact that the object emitting light is moving towards the observer. For example, light from the Andromeda Galaxy is blue-shifted because the Andromeda Galaxy is moving towards our own Milky Way Galaxy within the Local Group of galaxies. Blueshift also describes the shortening of wavelengths outside the visible spectrum – for example, of radio waves and X-rays – because their source is moving towards the observer.

ellipse A flattened circle. The shape is created by a point moving in a closed curve in which the sum of the point's distance from two fixed points is always constant. Orbits in space – of a satellite around its primary, the planets of the solar system around the Sun, and of stars around one another or the centre of a galaxy – are elliptical.

gravity Force that attracts physical bodies towards one another. In space, gravity acts with a strength proportional to the product of the mass of two bodies and inversely proportional to the square of the distance separating them. On Earth, gravity causes an object to have weight and to drop to the ground when released. In space, gravity has many effects – for example, keeping the Earth and the other planets in orbit around the Sun or keeping the Sun in orbit around the centre of the Milky Way Galaxy. Gravity is also the force that creates a black hole, when in a region matter becomes so compressed and mass so large that gravity is sufficiently powerful to draw everything in the vicinity into the black hole.

Local Group Group of more than 30 galaxies, with a diameter of 10 million light-years, that contains our own galaxy (the Milky Way), our nearest spiral galaxy (the Andromeda Galaxy) and the Triangulum Galaxy. The Group's gravitational centre lies between the Milky Way and the Andromeda Galaxy.

Proxima Centauri Red dwarf star in the constellation Centaurus, the closest star to our own Sun at a distance of around 4.3 light-years. The two stars of Alpha Centauri, our second and third nearest stars, are only 0.2 light-years from Proxima Centauri. At one time, astronomers classified Alpha Centauri as a binary star system, but they now generally view Proxima Centauri and the two stars of Alpha Centauri as a single entity, a triple star system.

redshift The stretching of light's wavelength towards the red end of the spectrum, caused by the fact that the object emitting light is moving away relative to the observer. Light from distant galaxies that are moving away from us is 'redshifted'. As with blueshift (see above), redshift also describes the lengthening of wavelengths outside the visible spectrum because their source is moving away from the observer.

spacetime Continuum of space and time, proposed by German-Swiss-American theoretical physicist Albert Einstein. While, in the conventional view of the Universe, the three dimensions of space and the single dimension of time were viewed as separate, spacetime is a four-dimensional continuum.

LIGHT-YEARS & PARSECS

the 30-second astronomy

Before 1543, most astronomers argued that the Earth does not move because the stars do not change position. In that year, Nicolaus Copernicus showed that the Earth does move – it orbits the Sun; since stars do not move relative to each other, it followed that stars must be at great distances. In fact, stars do appear to move by tiny amounts; the apparent movement of a star due to the Earth's motion is called 'parallax'. The first astronomer to measure a star's parallax was Friedrich Bessel, who in 1838 measured the parallax of the star 61 Cygni to be ⅓ of 1 second of arc (⅓ of 1 second of arc is approximately 1/10,000 of a degree), roughly as thick as a needle seen from 135 metres (150 yards). The distance of a hypothetical star whose parallax is 1 second of arc is defined as 1 parsec – equivalent to 30.6 trillion kilometres (19 trillion miles) – although the nearest star is not quite this close. Bessel illustrated the distance of 61 Cygni by calculating that light travel time from the star is 11 years. A 'light-year' – the distance light travels in one year – is 9.5 trillion kilometres (5.9 trillion miles).

3-SECOND BANG

A star's distance from Earth is expressed as the time light takes to travel to us, and its shift of position as Earth orbits Sun.

3-MINUTE ORBIT

Conjuring up the stars' distance in seventeenth-century terms, the natural philosopher Francis Robartes wrote in 1694 that 'Light takes up more time in Travelling from the Stars to us, than we in making a West-India Voyage (which is ordinarily performed in six Weeks).' The nearest star, Proxima Centauri, is at a distance of 4.3 light-years – 40.9 trillion kilometres (25.4 trillion miles) – and has a parallax of 0.7 of a second of arc.

RELATED TOPICS

See also
COLOUR & BRIGHTNESS OF STARS
page 54

3-SECOND BIOGRAPHIES

OLE RØMER
1644–1710
Danish astronomer who first measured the speed of light

FRIEDRICH BESSEL
1784–1846
Prussian mathematician and astronomer

30-SECOND TEXT

Paul Murdin

Light travels from the nearest star to the Earth in 4 years, but from the Sun to the Earth in only 8 minutes.

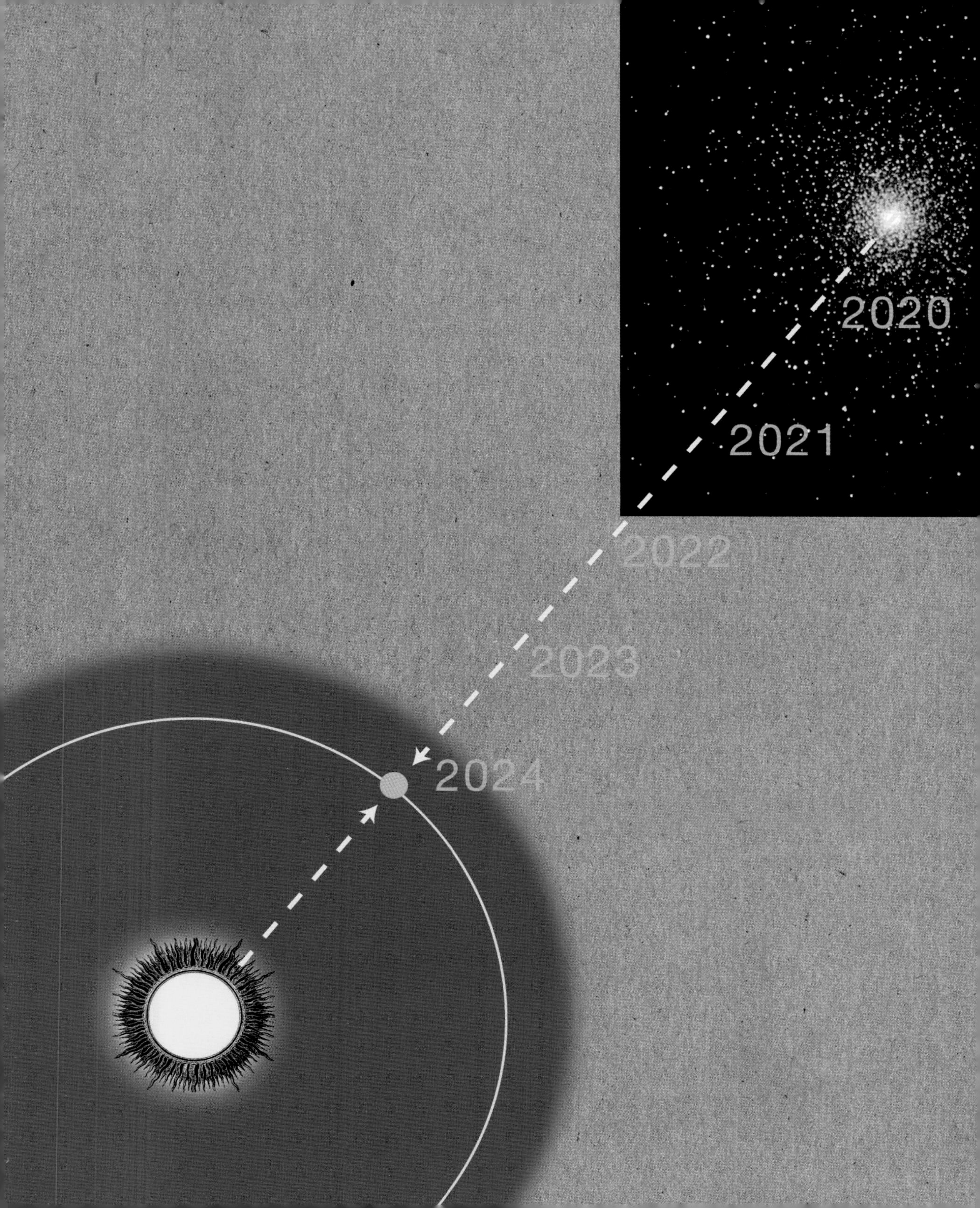
2020
2021
2022
2023
2024

ELLIPSES & ORBITS

the 30-second astronomy

3-SECOND BANG
Gravity is the most important force determining the motions of celestial bodies, and Isaac Newton's discovery of how it works launched the scientific era.

3-MINUTE ORBIT
An ellipse is a good approximation to the orbit of a planet in the solar system because the gravitational attraction of the Sun is dominant and that of the other planets negligible. In the long run, however, the orbits of two or more planets around a star are chaotic, the planets deflecting each other and looping in unrepeating orbits, which eventually become incalculable. Gravitational orbits are not as deterministic as is often thought.

Astronomers up to the sixteenth century believed that the planets moved around the Sun in circles, or combinations of circles called epicycles. Tycho Brahe set up a pre-telescopic observatory to measure how they really moved, and in 1605 his pupil, Johannes Kepler, used Brahe's measurements to show that the planets actually orbit the Sun in ellipses. Why they do this was a mystery, until Isaac Newton demonstrated that it was a consequence of his theory of gravitation; in particular, that the force of gravity between the Sun and a planet was proportional to the inverse square of the distance separating them. The orbit of one star around another or of a satellite around its parent planet are also elliptical. Newton took pride in the fact that his theory was 'universal' – and it was a further triumph when Newton's friend, Edmond Halley, used the theory to show that the orbit of a particular comet around the Sun was a flattened ellipse, and that it would return in 74 years. (We now know the comet as 'Halley's Comet'.) More typical comets orbit the Sun in a parabola; this geometric figure can be looked at as an extreme ellipse, very long and thin.

RELATED TOPICS
See also
BINARY STARS
page 56

EXOPLANETS
page 142

3-SECOND BIOGRAPHIES
TYCHO BRAHE
1546–1601
Danish astronomer

JOHANNES KEPLER
1571–1630
German discoverer of three laws of planetary motion

EDMOND HALLEY
1656–1742
English astronomer

EDWARD LORENZ
1917–2008
American discoverer of the theory of chaos

30-SECOND TEXT
Paul Murdin

The planets and the asteroids orbit the Sun in ellipses. Some comets orbit in ellipses, some in parabolas.

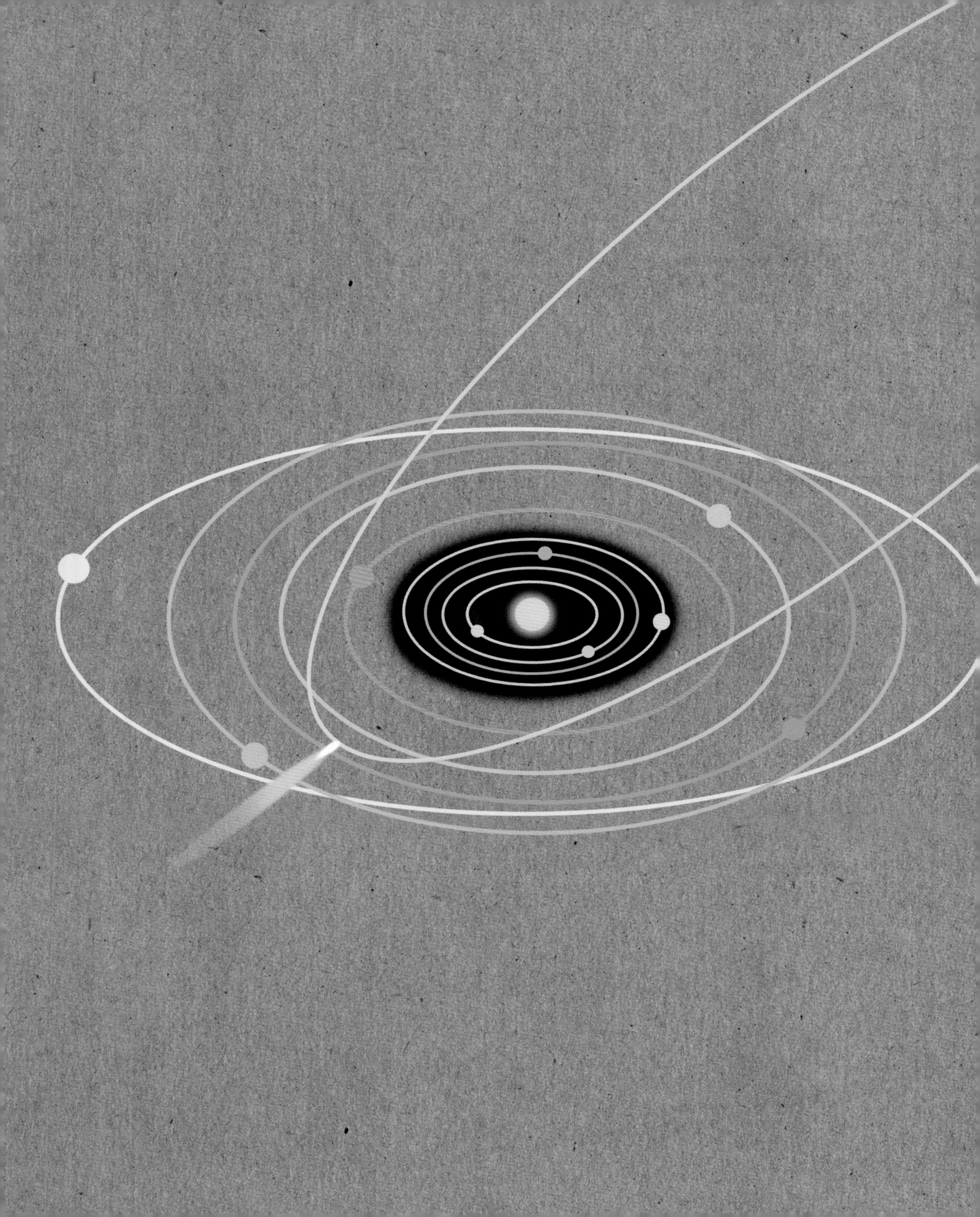

THE LIGHT SPECTRUM

the 30-second astronomy

Light travels as waves and the wavelength, or size, of the waves determines their colour. All light we see is a combination of many wavelengths of visible light, from 400 to 750 nanometres in wavelength, or from blue/violet to red. Astronomers use 'spectrometers' to split light from an object into a rainbow and measure its spectrum, its brightness at each individual wavelength. Our eyes are spectrometers, too, but coarse ones. They lump the multitude of wavelengths of a spectrum into three broad groups, so we perceive colours simply as a mixture of 'red', 'green' and 'blue'. Incandescent lightbulbs (which have a smooth distribution of wavelengths) and fluorescent bulbs (which have only a few distinct wavelengths) appear the same to our eyes, although their spectra are different when viewed through spectrometers with higher resolution. Spectrometers enable astronomers to analyze distant objects without visiting them. Different atoms and molecules emit or absorb in different sets of wavelengths; by observing these spectroscopic fingerprints, astronomers can determine the mineralogy of asteroids, the composition of stars, the gravity of white dwarfs, the motions of galaxies, the dynamics of accreting black holes, and more – all from the comfort of a telescope control room.

3-SECOND BANG
There is more to light than meets the eye! Astronomers explore the cosmos by splitting light from celestial objects into millions of colours.

3-MINUTE ORBIT
Light waves are similar to sound waves. When a car with a siren comes towards you and zooms past, you hear the pitch of the siren change – the car's motion is compressing and stretching the sound waves. Likewise, the wavelengths of light from objects can be blueshifted (compressed) or redshifted (stretched) due to motion. High resolution spectroscopy can detect motions even as minuscule as 1 metre (40 inches) per second, like those from planets tugging on their stars.

RELATED TOPICS
See also
THE SUN
page 36

COLOUR & BRIGHTNESS OF STARS
page 54

BEYOND VISIBLE LIGHT
page 102

3-SECOND BIOGRAPHIES
ISAAC NEWTON
1642–1727
English physicist

CECILIA PAYNE-GAPOSCHKIN
1900–79
English-American astrophysicist

30-SECOND TEXT
Zachory K. Berta

The spectrum of starlight is mostly a smooth, continuous rainbow of light, but it is missing light at very specific wavelengths, due to absorption by the star's atoms and molecules.

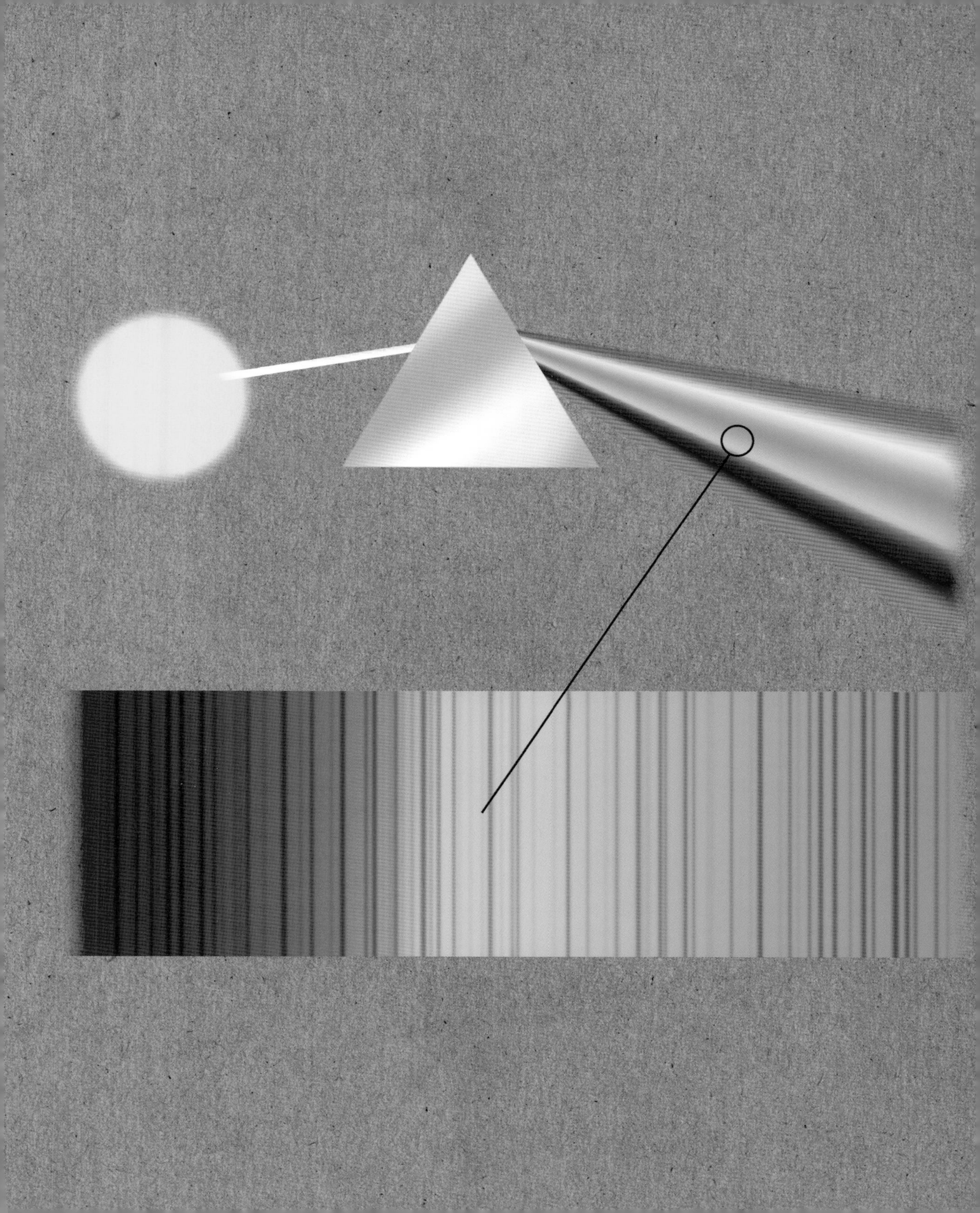

GRAVITY

the 30-second astronomy

In the seventeenth century, English physicist and mathematician Isaac Newton introduced our basic view of gravity as an invisible pull acting on an object from a distance. His law of gravity states that each body in the Universe exerts a force on any other: a stronger pull is produced by larger masses, and its strength drops with their increasing separation. Gravity gives a mass its weight, and dictates in which direction a supported object moves when released. The simple principle of gravity accounts for much of the observed behaviour of the nearby Universe. It accounts precisely for almost all the orbital motion of the planets and their moons, enabling space agencies successfully to send robotic probes to explore the solar system. Gravity dictates the motion of the stars in our galaxy, and the galaxies within clusters, such as our Local Group. Our understanding of gravity has been enhanced by German-Swiss-American physicist Albert Einstein's General Theory of Relativity (1915), which supersedes Newton's ideas when the velocities involved approach the speed of light. We understand more about what gravity does than what it is – unifying gravity with quantum theory remains a major unsolved problem in physics.

3-SECOND BANG
Gravity is fundamental to our understanding of the Universe because it is the force that dictates the motion and interactions of all astronomical bodies.

3-MINUTE ORBIT
Any change in the strength of the gravitational pull experienced across a cosmic object creates tidal forces. The large difference in force experienced by Earth's oceans both closest to and furthest from the Moon creates two high tides a day. Tidal forces between pairs of merging galaxies rip out long streams of stars and gas; and stars passing too close to a black hole can be completely shredded by the tidal forces.

RELATED TOPICS
See also
THE MOON
page 20

GALACTIC STRUCTURES
page 88

ELLIPSES & ORBITS
page 120

RELATIVITY
page 126

3-SECOND BIOGRAPHIES
ISAAC NEWTON
1642–1727
English physicist

ALBERT EINSTEIN
1879–1955
German-Swiss-American theoretical physicist

30-SECOND TEXT
Andy Fabian

When galaxies pass close to each other at relatively slow speeds, the gravitational pull of one on another tears out 'tidal' tails of gas and stars, distorting their spiral shape.

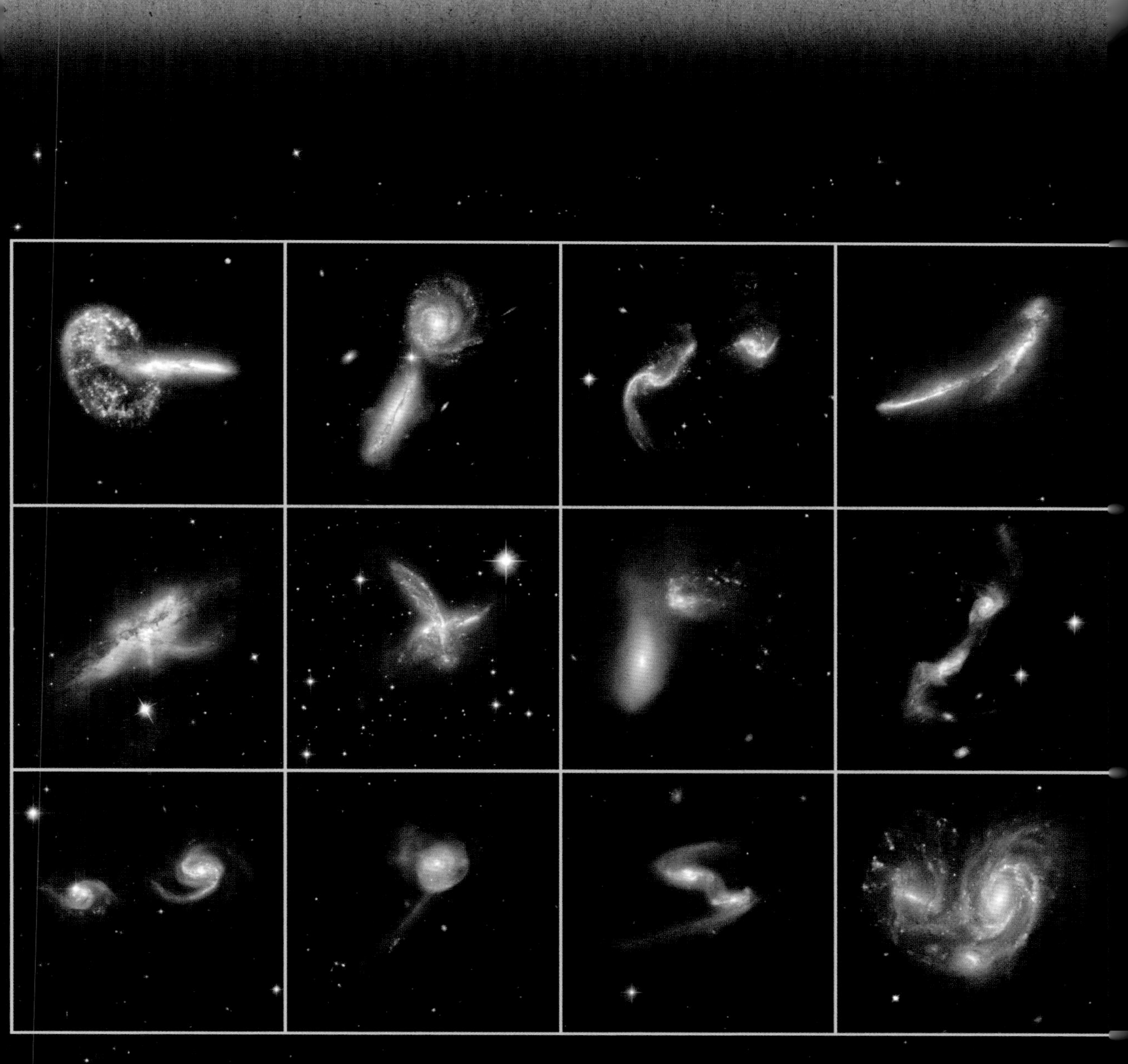

RELATIVITY

the 30-second astronomy

In his Special Theory of Relativity (1905), Albert Einstein stated that measurements of length and time intervals are affected by the relative speeds of a person's observations and the event or object being observed. Observations by a stationary observer of a rapidly moving clock show it to run slow, and measurable lengths are observed to be shorter than when remaining at rest. Despite this, the speed of light and the laws of physics remain unchanged for all observers, no matter how fast they move. This leads to Einstein's famous equation, $E=mc^2$, which expresses how any mass can also embody energy – this concept explains how the mass released through atomic reactions gives stars energy. In 1915, Einstein extended his ideas to a General Theory of Relativity, which encompassed accelerations between the observer and object being observed. This gave a new account of gravity as being caused by the curvature of space and time in the presence of a mass. Thus the distribution of matter in the Universe influences the overall shape of space. Prominent also is the principle of equivalence, which states that it is impossible on small scales to distinguish between the downward pull of gravity and upward acceleration of an observer.

3-SECOND BANG
Relativity describes how the relative speed of an observer and object, and changes in speed, alter measurements of distance and time, and the workings of gravity.

3-MINUTE ORBIT
The predictions of relativity have so far been proved correct by each experiment designed to test them. Unlike Newton's theory of gravity, relativity can account for anomalies in Mercury's orbit, mirages caused by gravitational lensing, and the slowing of time in a stronger gravitational field. It also explains the spiralling-in of very close binary stars, due to the emission of gravitational waves – ripples of spacetime propagating outwards at the speed of light.

RELATED TOPICS
See also
BINARY STARS
page 56

BLACK HOLES
page 70

GRAVITY
page 124

GRAVITATIONAL LENSING
page 128

3-SECOND BIOGRAPHIES
HENDRIK LORENTZ
1853–1928
Dutch physicist

ALBERT EINSTEIN
1879–1955
German-Swiss-American theoretical physicist

30-SECOND TEXT
Andy Fabian

The shape of space is curved into deep wells around massive objects such as planets, stars and galaxies.

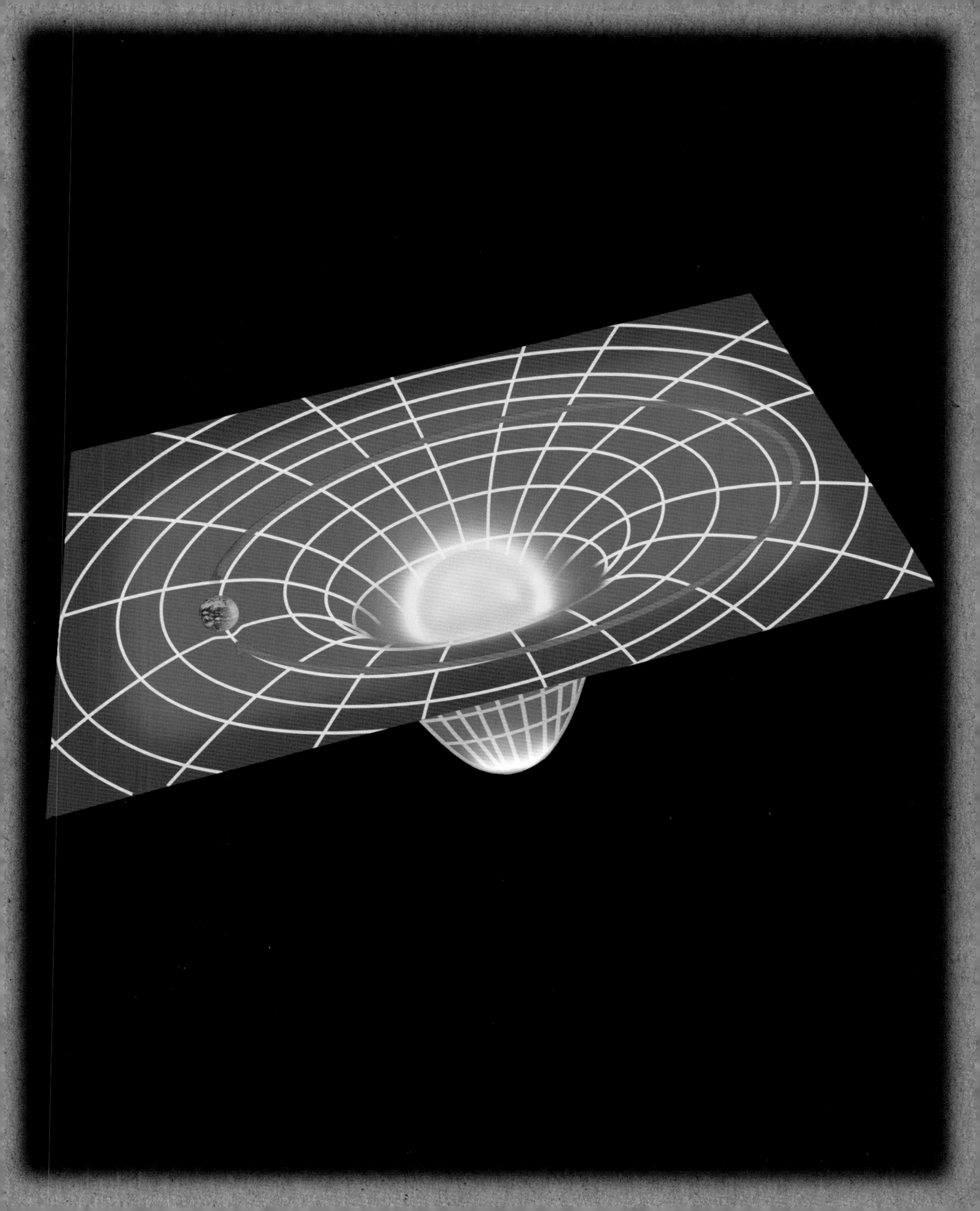

GRAVITATIONAL LENSING

the 30-second astronomy

Magnifying lenses on Earth work by focusing light that passes through them, using refraction to bend light as it passes between air and glass. Out in the vacuum, light travels along straight paths through space, never deviating from its initial trajectory. But what happens if light passes by a very massive object? The strong gravity associated with the object's mass will – according to Albert Einstein's General Theory of Relativity – cause space itself to bend; if space bends, light travelling through it will appear to bend, too. Galaxy clusters, weighing 1,000,000,000,000,000 times the mass of our Sun, can act as powerful 'gravitational lenses', magnifying light from background galaxies, sometimes distorting their appearance into beautiful, slender arcs on the sky. A few rare gravitational lenses happen to be positioned just perfectly to provide a natural (if immovable) zoom lens in front of our telescopes, allowing astronomers to peer at remarkably detailed features of young galaxies still in the early stages of formation out at the edge of the observable Universe. In addition to galaxy clusters, many other astronomical objects can act as gravitational lenses on smaller scales, from powerful supermassive black holes all the way down to tiny little planets.

3-SECOND BANG

The Universe is scattered with gravitational lenses, enormous astrophysical magnifying glasses that focus or distort background starlight long before it reaches our telescopes.

3-MINUTE ORBIT

Astronomers monitoring the brightness of millions of stars have found rare sparkles of light caused by gravitational 'microlensing' from stars and/or planets in our galaxy. Stars and planets drifting through space sometimes align with distant background stars, catch their light, focus it to Earth, and dramatically magnify them for a brief time. To date, more than a dozen planets have been discovered with the microlensing magnification technique.

RELATED TOPICS

See also
BLACK HOLES
page 70

DARK MATTER
page 110

GRAVITY
page 124

RELATIVITY
page 126

3-SECOND BIOGRAPHIES

FRITZ ZWICKY
1898–1974
Swiss astronomer

BOHDAN PACZYNSKI
1940–2007
Polish astronomer and leading researcher into gravitational lensing and microlensing

30-SECOND TEXT

Zachory K. Berta

A massive galaxy cluster bends light from background objects, creating multiple, magnified images of them that we can observe with our telescopes.

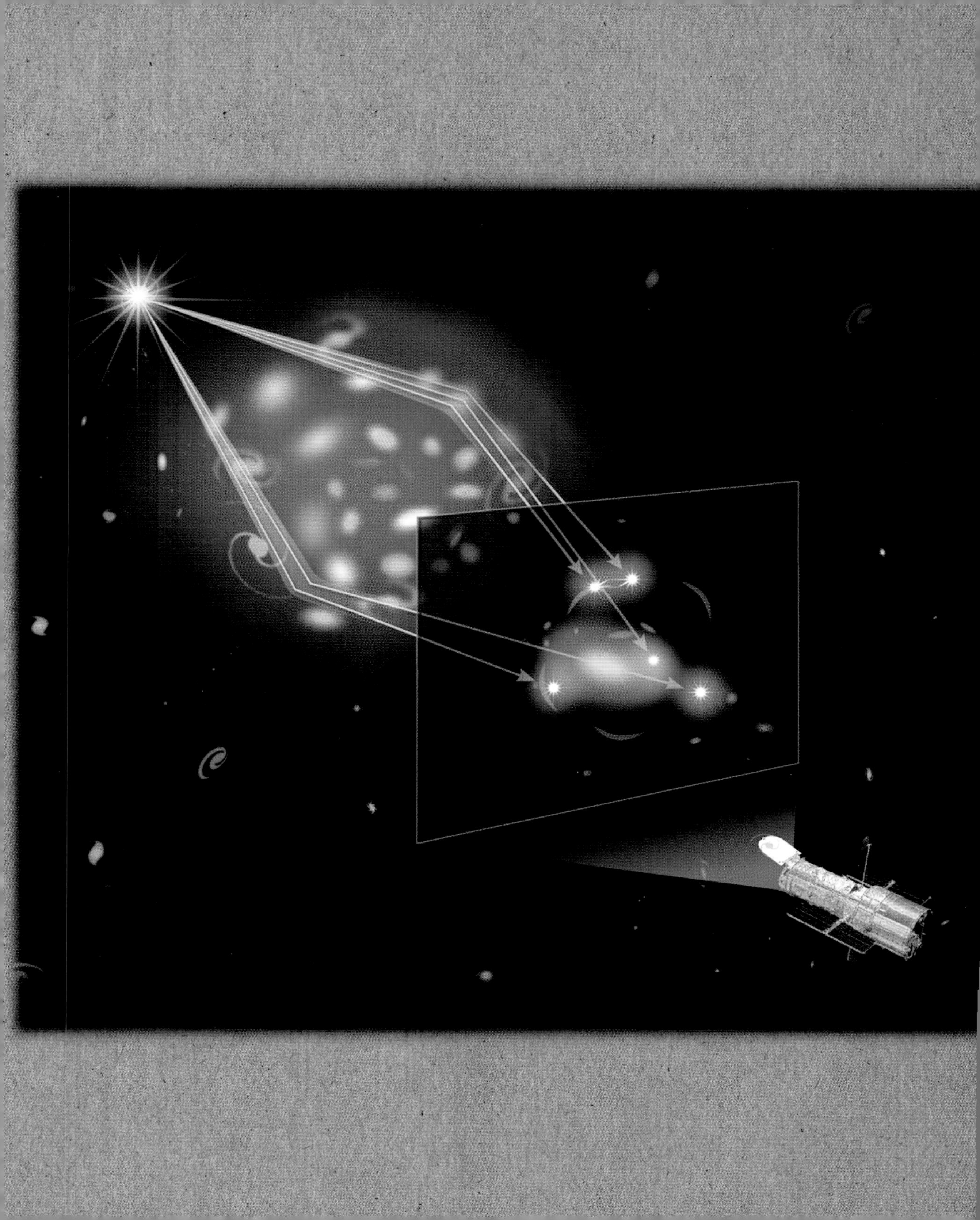

14 February 1898
Born in Varna, Bulgaria

1904
Sent to Switzerland for education; eventually studied in Zürich at the Swiss Federal Institute of Technology, ETHZ

1925
Emigrated to the USA to work at Caltech on Rockefeller fellowship

1933
Inferred the existence of dark matter

1934
Coined term 'supernova'; published (with Walter Baade) *Cosmic Rays from Super-Novae*

1935
Pioneered (with Baade) use of Schmidt telescope

1937
Posited that galaxy clusters and nebulae could act as gravitational lenses, as predicted by Einstein

1942
Appointed professor of astronomy at Caltech

1943–1961
Research consultant/ Director at Aerojet Engineering Corporation

1946
Published *On the Possibility of Earth-Launched Meteors*

1949
Awarded Presidential Medal of Freedom for work on rocket propulsion

1961–1968
With colleagues compiled six-volume *Catalog of Galaxies and Clusters of Galaxies.* Appointed Professor Emeritus at Caltech (1968)

1969
Published *Discovery, Invention, Research through Morphological Analysis*

1971
Self-published *Catalog of Selected Compact Galaxies*

1972
Awarded Gold Medal of the Royal Astronomical Society

8 February 1974
Died in Pasadena; buried at Mollis, Switzerland

FRITZ ZWICKY

Best known as the father of dark matter, which he identified and named, Fritz Zwicky was born in Bulgaria to a Swiss father and a Czech mother, educated at the prestigious Eidgenössische Technische Hochschule in Zürich and spent all his working life in the United States, most of it at the California Institute of Technology. He became Caltech's (and the world's) first astrophysicist, simply by deciding to marry his physics training with astronomy. He was a maverick and original thinker, many of whose early ideas and theories, mocked by some contemporaries, have developed into orthodoxy: the existence of dark matter, neutron stars, gravitational lenses, and supernovae; those that didn't (nuclear goblins, shifting and/or customizing the solar system, creating artificial meteors) still loiter in the realms of science fiction. On the practical side is Zwicky's indispensable six-volume catalogue of galaxies and his pioneering work on jet propulsion during and after the Second World War.

Zwicky did not just think outside the box, he developed his own box for thinking in. This was morphological analysis (MA), the refinement of a scientific investigative tool introduced by Goethe. It works by aggregating all data, however apparently unrelated, into a matrix, and looking at all possible outcomes to a problem, even the most startling solutions. An apocryphal story has him ordering an assistant to fire a revolver through the slit of a telescope to clear turbulence; it didn't work, but is a fine example of his radical, left-field turn of mind.

Zwicky's glittering scientific reputation has been rather compromised by his attitude to peers and students; although a principled humanitarian, he was not at his best around individuals; he was famously unable to suffer fools at all, let alone gladly, and was convinced that he was always right and everyone else was stupid – even Robert Oppenheimer. At his death, he was still locked in confrontation with some of his students over an excoriating introduction he wrote to his *Catalog of Selected Compact Galaxies (1971)*, in which he styled himself as a heroic lone wolf and all others as scatter-brained, date-bending sycophants and fawning apple-polishers.

WORMHOLES

the 30-second astronomy

Albert Einstein's General Theory of Relativity allows for the existence of black holes that create a bridge to a different place in the Universe, or to another universe. This bridge takes the form of a tube (wormhole) that links separate points in spacetime. If space is simplified as a two-dimensional sheet curved back on itself, a wormhole can be visualized as a hollow tunnel making a 'shortcut' between the two layers. The time taken to pass through a wormhole would be far less than a journey on a path through normal space, and so the wormhole could potentially provide a means of faster-than-light travel. If one mouth of the wormhole were greatly accelerated and the other kept stationary, then the stationary mouth would age less rapidly according to the theory of time dilation in Einstein's Special Theory of Relativity. Thus it might be possible to crawl into the moving end of the wormhole only to emerge prior to one's time of entry, and thus travel through time. However, wormholes are so far a purely theoretical concept, and do not arise naturally from the type of black hole created by the collapse of a massive star.

3-SECOND BANG

A wormhole is a hypothetical tunnel connecting different regions of space and time – either within our own universe, or leading to a parallel universe.

3-MINUTE ORBIT

Wormholes are predicted to be unstable, limiting their potential to provide a means for space and time travel. They would collapse so rapidly that no matter could make a journey through before the exit route was pinched off. Some scientists suggest that the mouth of a wormhole could be kept open were it filled with exotic matter that produced 'antigravity', such as that speculated to fulfil the role of dark energy in the Universe.

RELATED TOPICS

See also
BLACK HOLES
page 70

DARK ENERGY
page 112

RELATIVITY
page 126

3-SECOND BIOGRAPHIES

ALBERT EINSTEIN
1879–1955
German-Swiss-American theoretical physicist

NATHAN ROSEN
1909–95
American-Israeli physicist

JOHN ARCHIBALD WHEELER
1911–2008
American theoretical physicist

30-SECOND TEXT

Andy Fabian

A wormhole can be visualized as a potential connection between different times and places in the Universe.

OTHER WORLDS

OTHER WORLDS
GLOSSARY

51 Pegasi Star in the constellation Pegasus, 50.9 light-years from the Earth. It is orbited by the planet 51 Pegasi B, and was the first exoplanet to be identified.

astrobiology The development and testing of hypotheses concerning extraterrestrial life. Astrobiologists also research into the early development of life on Earth. Exobiology refers to the narrower study of the effects of extra-terrestrial environments on hypothetical beings and the implications for extraterrestrial life.

astrophysics Study of the physics of the Universe, incorporating investigation of celestial objects' physical properties and interactions.

carbon Element found in all known life forms. In the Universe, it is the fourth most abundant element after hydrogen, helium and oxygen. It is the product of the burning of helium in nuclear fusion reactions within stars.

Curiosity NASA vehicle that landed on Mars on 6 August 2012, part of the NASA Mars Science Laboratory launched on 26 November 2011.

exoplanets Planets outside the solar system.

extraterrestrial life Any form of life that does not have its origin on Earth. The term can refer to any hypothetical life-form, from an organism consisting of a few cells to a more conventionally imagined 'alien'.

gaseous envelope Cloud of gases, held together by gravity. An envelope of gases forms a nebula. The Earth's atmosphere is also sometimes called a gaseous envelope.

geophysics Study of the physics of the Earth and its atmosphere.

GJ 1214b An exoplanet orbiting the star GJ 1214, around 40 light-years from the Sun in the constellation Ophiuchus. GJ 1214b, discovered in 2009, is an example of a super-Earth, with a radius and mass larger than Earth but smaller than the gas giants of our solar system. Planet GJ 1214b is considered a likely candidate to be an ocean planet (one whose surface is entirely covered by water). Observations of GJ 1214b with the Hubble Space Telescope appear to indicate that water accounts for a major part of its mass.

HD 209458b A exoplanet orbiting the star HD 209458, around 150 light-years from the Sun in the constellation Pegasus. It was the first exoplanet to be observed in transit to its star, and the first exoplanet to have its atmosphere studied. Its orbit is so close to its star that it has a year lasting only 3.5 Earth days and a surface temperature estimated at 1,000°C (1,800°F).

Kepler 10b An exoplanet in orbit around the star Kepler 10, some 564 light-years away from the Sun in the constellation Draco. The NASA Kepler mission targeted Kepler 10 and discovered a planetary system containing at least two planets, named Kepler 10b and Kepler 10c.

Kepler mission NASA discovery mission number 10, a space satellite launched to detect super-Earth exoplanets in the habitable zone of their stars (area in which planets might possibly have sufficient atmospheric pressure to have liquid water on their surface, and so might harbour life).

macroscopic As opposed to microscopic. A macroscopic object can be seen with the naked eye, while a microscopic object can be seen only using a microscope.

ocean planets Hypothetical planets whose surface is entirely covered by ocean. Astronomers think it likely that some planets migrate inward towards their star during the process of formation, and that as a result an icy planet might become an ocean planet when the ice melted to water.

perturbation Movement of a planet or large body due to forces that do not originate from the gravitational attraction of a single other source. Perturbation can be the effect of planets or natural satellites, or of an atmosphere – any source other than the one that controls the body's normal motion. Astronomers can identify exoplanets, for example, by detecting gravitational perturbations in the orbit of a star. The normal orbit around the centre of the galaxy is being disturbed by the gravitational effect of the planet; although the exoplanet is too far away to see, the astronomers can deduce its existence from the perturbation of the star.

protoplanetary disc Spinning disc of gas surrounding a newly formed star.

super-Earths Exoplanets with mass between those of the Earth (mass 5.9722 x 1024 kg) and Neptune (102.4 x 1024 kg). Neptune's mass is 17.5 times that of Earth.

transit Passing of a planet in front of its star. During a transit astronomers using the Hubble Space Telescope can take readings indicative of the planet's atmosphere.

Viking NASA discovery mission to Mars. Two spacecraft, *Viking 1* and *Viking 2*, launched in August and September 1975 and landed on Mars in July and September 1976.

EXTRATERRESTRIALS

the 30-second astronomy

3-SECOND BANG
Whether extraterrestrial life exists is a defining question for humanity, which remains unanswered despite recent indications that life might not be uncommon in the Universe.

3-MINUTE ORBIT
It is difficult to predict what alien life could look like. Life is usually defined as a system that is self-organizing, reproducing, responsive to its environment and evolving in successive generations. The easiest way to accomplish this seems to be based on the chemistry of the carbon atom in liquid water, but this idea is, of course, inspired by life as we know it.

The more humanity has learned about space in the many centuries since ancient astronomers placed the Earth at the centre of the Universe, the more we have realized that there is nothing unusual about the solar system. The most recent milestone in this process of discovery is the identification of exoplanets, and hints that Earthlike planets are extremely common. These new insights could indicate that life as we know it, based on common atoms and molecules, may be an ordinary thing in the Universe. However, humanity has still not recorded any indication that life exists elsewhere, either within or outside our solar system, nor found any sign that our Earth has been visited by another life form. This simple fact indicates that highly advanced extraterrestrial civilizations with the desire to communicate are rare, which may appear as a paradox considering the probably extremely large number of 'other Earths' and the fact that technological evolution seems exponential. A possible explanation to this paradox is that the lifespan of advanced civilizations is short, a hypothesis that has a strong echo at a time where the Earth's human population is realizing the alarming rate at which it is changing its own environment.

RELATED TOPICS
See also
THE EARTH
page 18

EXOPLANETS
page 142

TOWARDS ANOTHER EARTH
page 148

3-SECOND BIOGRAPHIES
ENRICO FERMI
1901–54
Italian-born American physicist

FRANK DRAKE
1930–
American astronomer and astrophysicist

JILL TARTER
1944–
American astronomer and astrobiologist

30-SECOND TEXT
François Fressin

Life on Earth is based on the DNA molecule. Would extraterrestrial life also develop from macromolecule replication?

9 November 1934
Born in Brooklyn, New York

1954
Graduated from the University of Chicago as Bachelor of Arts

1955
Bachelor of Science in Physics at the University of Chicago (and Masters in 1956)

1960
PhD in Astronomy and Astrophysics at the University of Chicago

1960–62
Miller Fellow at the University of California

1962
NASA space probe *Mariner 2* supports his hypothesis that the surface of Venus is very dry and hot

1962–68
Worked at the Smithsonian Astrophysical Observatory in Cambridge, Mass, and lectured at Harvard

1971
Worked at Cornell University, Ithaca, New York

1972
Became full professor at Cornell, Director of Planetary Studies, and (until 1981) Associate Director of the Center for Radio Physics and Space Research

1972
NASA space probe *Pioneer 10* launched, carrying Sagan-designed communication plaque

1977
Awarded NASA's Distinguished Public Service medal

1978
Won Pulitzer Prize for nonfiction with *The Dragons of Eden: Speculations on the Evolution of Human Intelligence* (1977)

1979
Wrote *Broca's Brain: Reflections on the Romance of Science*

1980
Co-wrote and narrated award-winning TV series *Cosmos: A Personal Voyage*

1982
Organized petition advocating the establishment of the SETI Institute (Search for Extraterrestrial Intelligence) in the journal *Science*

1984
SETI Institute set up; became a member of the Board of Trustees

1985
Wrote *Contact*, later made into a movie (1997)

1990
Awarded Oersted medal

1994
Awarded the National Academy of Science's Public Welfare Medal

1995
Wrote *Pale Blue Dot: A Vision of the Human Future in Space*

1996
Wrote *The Demon-Haunted World: Science as a Candle in the Dark*

20 December 1996
Died in Seattle

1997
Billions and Billions: Thoughts on Life and Death at the Brink of the Millennium published posthumously

CARL SAGAN

Astronomer, astrophysicist, cosmologist and prolific writer, Carl Sagan was an unashamed popularizer of what he loved best, and he enjoyed a high public profile (to the annoyance and even chagrin of some of his peers). Captivated by stars at the age of five, according to his own memoirs, he combined a sense of awe at the cosmos with an unswerving commitment to the rigours of the scientific method: keep your mind open, but constantly test what comes into it. He devised the 'Baloney Detection Kit', a set of mental tools to help anyone with a lively mind debunk junk science and disrobe charlatans, and urged its use at all times.

Sagan actively embraced the popular media, producing an internationally successful TV series in the 1980s, explaining the cosmos as we then knew it, and wrote or worked on more than 20 books aimed at the nonscientific reader. At the same time, he maintained a successful 'hard-science' career – at Cornell University, New York, in the main. It was perhaps lucky for him that his career took off at about the same time as the NASA space programme; he was an adviser throughout, from the 1950s (when he was still a PhD student) onwards, briefing the Apollo astronauts and designing experiments to be carried out by robotic spacecraft. In his astronomer's chair, he theorized (and was proved largely correct) about the surface temperatures of Jupiter and Venus, seasonal changes on Mars, and the likelihood of water on Titan (a moon of Saturn) and Europa (a moon of Jupiter). He was also an early whistle-blower on the dangers of climate change, and (during the Cold War) on the catastrophic possibility of a nuclear winter, if the war ever turned hot.

Sagan is probably best known for his trail-blazing work on exobiology (the study of nonterrestrial biological conditions) and the search for extraterrestrial life. He encouraged the use of radio telescopes to detect signs of life, and devised the plaques sent out with the *Pioneer* and *Voyager* space probes designed to be decoded by intelligent life forms. True to his precepts, he applied stringent tests to all UFO sightings or unverifiable stories of alien abduction and concluded towards the end of his life that it was very unlikely that the Earth had ever been visited by extraterrestrial intelligence: but that did not mean to say it was not out there, somewhere.

EXOPLANETS

the 30-second astronomy

The existence of worlds orbiting stars other than the Sun has been imagined for centuries, but no such worlds were known to humanity until the end of the twentieth century. In 1995, Swiss astrophysicist Michel Mayor and his student Didier Queloz discovered the gravitational perturbation of an object as massive as Jupiter orbiting the star named 51 Pegasi. Since that date, astronomers have identified thousands of exoplanets. The diversity of planets they have found seems to be limited only by the technological possibilities of the telescopes used to find them. Some planets have been discovered orbiting binary stars; others are not bound to any star and are free-floating in space. There are scorched worlds with molten surfaces, giant planets with huge rocky cores several dozen times the mass of the Earth, planets darker than black paint, and water worlds that are probably covered completely with an ocean. On the path towards the discovery of a true analogue of the Earth a field of research has emerged. Combining astrophysics, geophysics and biology, the study of the habitability of exoplanets evaluates their suitability for hosting life.

3-SECOND BANG
Exoplanets, or extrasolar planets, are planets outside the solar system; their recent discovery intensified the search for other life forms in the Universe.

3-MINUTE ORBIT
Detecting 'other Earths' is extremely challenging. Our Earth is already practically invisible from our own probes in the outskirts of the solar system. However, astronomers can find the signature of an exoplanet by studying its gravitational influence on the motion of the star it orbits, or by detecting a temporary drop of a small amount of light due to a planet eclipsing a small part of a star.

RELATED TOPICS
See also
SUPER-EARTHS & OCEAN PLANETS
page 146

TOWARDS ANOTHER EARTH
page 148

3-SECOND BIOGRAPHIES
GIORDANO BRUNO
1548–1600
Italian astronomer burned at the stake by the Roman Inquisition for claiming the Universe may contain many other inhabited worlds

MICHEL MAYOR & DIDIER QUELOZ
1942– & 1966–
Swiss astrophysicists and pioneers of the search for exoplanets

30-SECOND TEXT
François Fressin

51 Pegasi was the first star known to host an exoplanet, a planet as large as Jupiter but orbiting it in just four days.

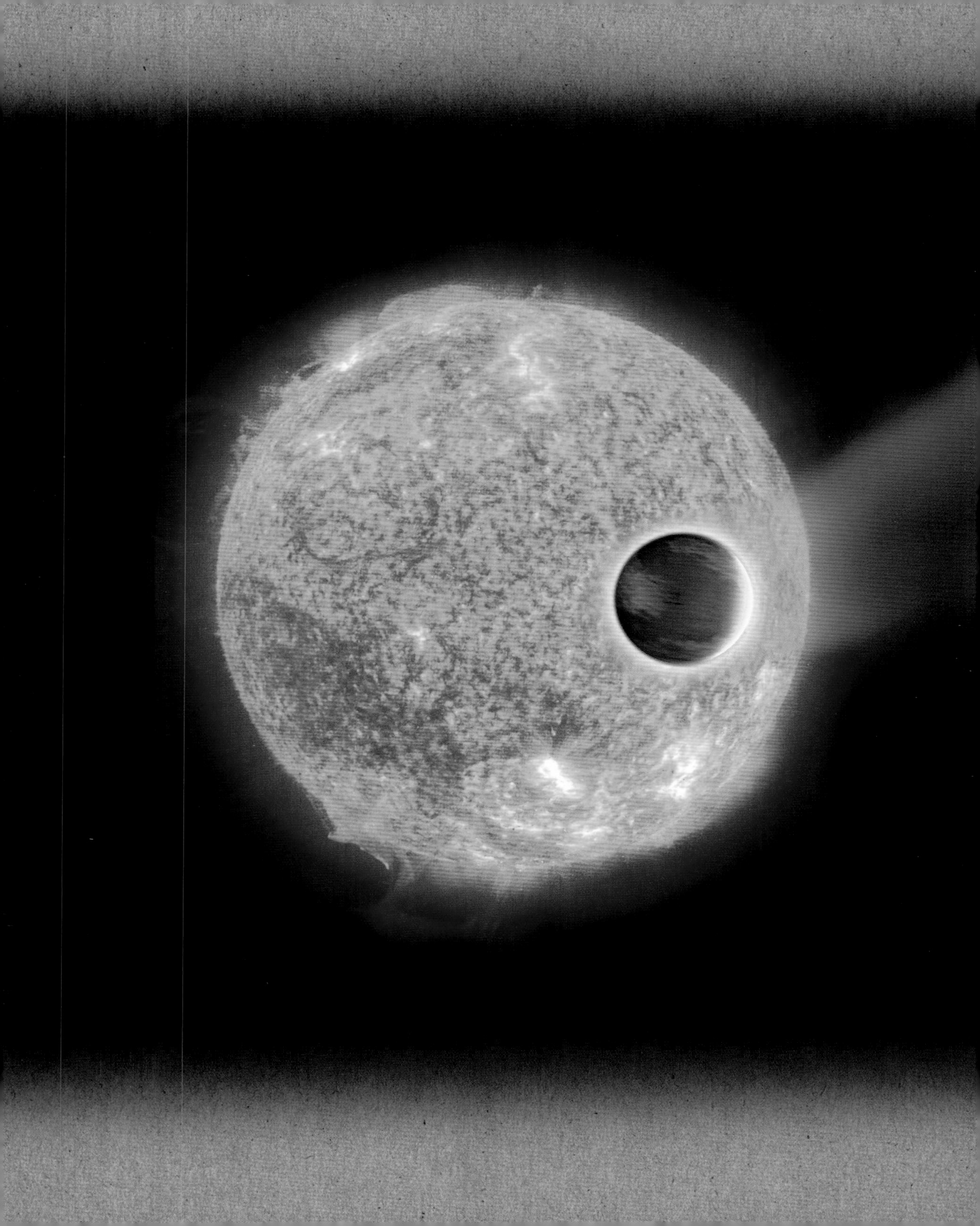

SUPER-EARTHS & OCEAN PLANETS

the 30-second astronomy

Of the planets that orbit our Sun, none have sizes between those of Earth and the ice-giant Neptune. Yet, because astronomers can determine the sizes of planets orbiting other stars by measuring how much starlight they block, we know that our galaxy seems to be teeming with planets in this size range. Although astronomers call these planets 'super-Earths' because of their size, many might be nothing like Earth. One such exoplanet is Kepler 10b, which is so much denser than Earth that it must be made entirely out of molten rock and iron. In contrast, the exoplanet GJ 1214b is far less dense than Earth and could be composed largely of steamy water vapour mixed with other gases – a water-world or ocean planet. Kepler 10b and GJ 1214b are both close to their host stars and very hot, but super-Earth planets in cooler orbits could have barren rock crusts, or puffy hydrogen atmospheres, or shifting continents like Earth, or maybe even worldwide oceans hundreds of miles deep. As astronomers study the masses, sizes and atmospheres of more super-Earths, we will learn more about the processes governing the formation and evolution of these exoplanets.

3-SECOND BANG

Exoplanets slightly bigger than the Earth could have a variety of compositions; astronomers are eager to unravel the mysteries of these new worlds of possibility.

3-MINUTE ORBIT

If a super-Earth grew a little too big in its youth, it would start a process called runaway accretion, quickly sweeping up huge swaths of gas from the protoplanetary disc and becoming a gas giant. It would be difficult for life to form on a planet with such a deep gaseous envelope. How big a planet can be and still be hospitable to biological life is an active area of research in exoplanetary science.

RELATED TOPICS

See also
THE EARTH
page 18

URANUS & NEPTUNE
page 30

EXOPLANETS
page 142

HOT JUPITERS
page 144

TOWARDS ANOTHER EARTH
page 148

3-SECOND BIOGRAPHIES

SARA SEAGER
1971–
Canadian-American astrophysicist and a leading researcher into super-Earth planets

30-SECOND TEXT

Zachory K. Berta

All we know about many recently discovered super-Earths is their size – bigger than our Earth but far smaller than planets such as Neptune or Jupiter.

TOWARDS ANOTHER EARTH

the 30-second astronomy

Recent technological advances have made it possible for astronomers to identify the first Earth-sized planets orbiting other stars, but we do not yet know how common these planets are, nor what fraction of them could sustain life. There are two techniques used to look for these faint and extremely distant objects – which are no more than tiny rocks orbiting fireballs one million times more massive than they are, but are seen together as a speck of light in telescopic images. The dynamical technique consists of identifying the motion of the planet around the star by looking for the reflex motion of the star pulled by the planet or the dimming of light when the planet eclipses a small part of the star during its orbit. The direct-imaging technique requires blocking the light coming from the star to see its surroundings, which otherwise would be invisible in the star's glare. Once telescopes have identified and gathered enough light, they can study the planet's atmospheric features and investigate how similar they are to those of the Earth. As the twenty-first century proceeds, we envisage imaging these planets and mapping these other worlds – and looking for seasonal changes and direct signs of life.

3-SECOND BANG
Astronomers now have the technology to find Earth-sized planets around other stars, with the goal of finding out how similar to the Earth they are.

3-MINUTE ORBIT
The Alpha Centauri system is the closest to the Sun, and by 2020 we could have proof that there is another Earth in that system. The next step could be to send a probe to take high-resolution pictures. Such a challenging project would probably involve several international generations of scientists by the time the probe reached Alpha Centauri. Running the project and sharing its results would be a unifying experience.

RELATED TOPICS
See also
THE EARTH
page 18

THE LIGHT SPECTRUM
page 122

EXOPLANETS
page 142

EVIDENCE FOR OTHER LIFE
page 150

3-SECOND BIOGRAPHIES
BERNARD LYOT
1897–1952
French astronomer

GEOFFREY MARCY
1954–
American astronomer, pioneer in the discovery of exoplanets

30-SECOND TEXT
François Fressin

Alpha Centauri A, the constellation Centaurus' brightest star, is the same type as our Sun, fuelling speculation that it might contain planets that harbour life.

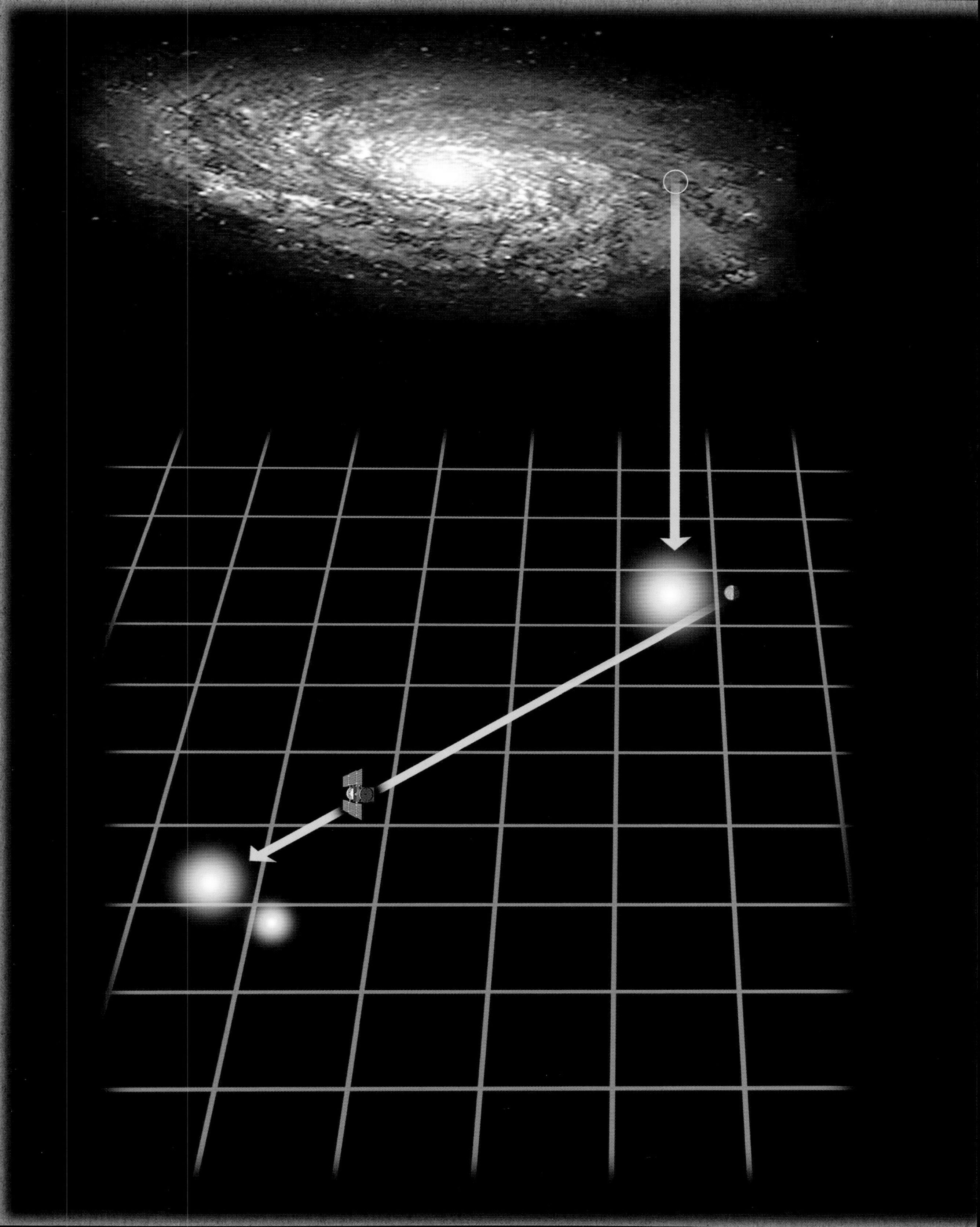

EVIDENCE FOR OTHER LIFE

the 30-second astronomy

The search for life on another planet can be considered to have started in the late nineteenth century, when observers thought they could see straight 'canals' on Mars that later proved to be optical illusions due to the instruments they were using. In the 1970s, the *Viking* spacecraft carried biological experiments to the surface of Mars but could not identify any conclusive sign of life. Despite many claims of unidentified flying objects (UFOs) visiting the Earth since, the only clear picture of a 'flying saucer' we have is one of the shell carrying the *Curiosity* rover that NASA sent to Mars in 2011–12, with the aim of studying the planet's habitability and whether it could have ever supported life. Astronomers are also investigating whether life could exist on other objects in the solar system, such as on Jupiter's moon Europa. However, our best chance of discovering another macroscopic life form is the investigation of 'other Earths' orbiting distant stars. By probing the atmospheric composition of such planets, astronomers search for molecules more likely to have been produced by living organisms (such as the oxygen and methane on Earth) than ordinary chemistry.

3-SECOND BANG
Astrobiology investigates whether life exists beyond Earth, and in what likely conditions, as well as how humans can detect it if it does.

3-MINUTE ORBIT
The search for extra-terrestrial intelligence aims at directly finding an advanced extraterrestrial civilization, or letting it find us. The most important effort involves the use of large radio telescopes to identify the fingerprint of another civilization's telecommunications, but such a signal has not yet been identified. Another question is how we could leave detectable fingerprints and what 'language' would be intelligible to an alien civilization.

RELATED TOPICS
See also
THE EARTH
page 18

EXTRATERRESTRIALS
page 138

TOWARDS ANOTHER EARTH
page 148

3-SECOND BIOGRAPHY
CARL SAGAN
1934–96
American astronomer, astrophysicist and writer

30-SECOND TEXT
François Fressin

Jupiter's satellite Europa shows cracks in its icy surface, indicative of a possible subsurface ocean where life could have developed.

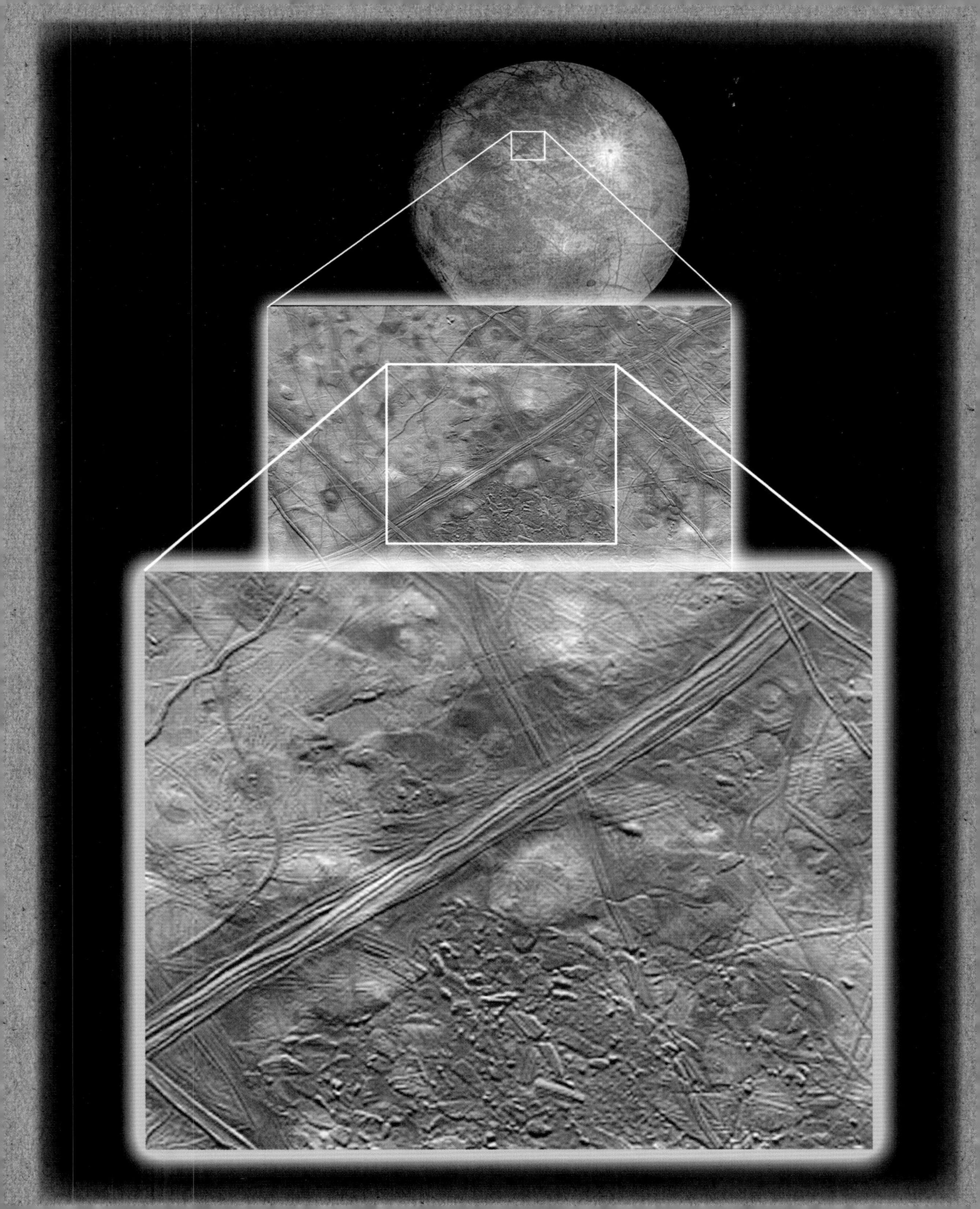

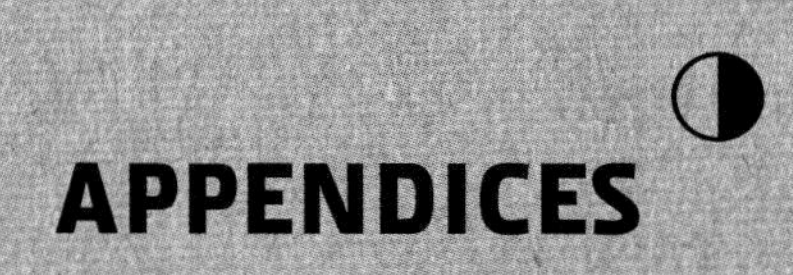
APPENDICES

RESOURCES

BOOKS

DK Illustrated Encyclopedia of the Universe
ed., Martin Rees
(Dorling Kindersley, 2011)

Exoplanets
Sara Seager
(University of Arizona Press, 2010)

Exploring the X-Ray Universe
Frederick D. Seward & Philip A. Charles
(Cambridge University Press, 2010)

Firefly Encyclopedia of Astronomy
Paul Murdin
(Firefly, 2004)

An Introduction to Modern Astrophysics
Bradley W. Carroll & Dale A. Ostlie
(Pearson, 2006)

Mapping the Universe
Paul Murdin
(Carlton Publishing, 2012)

Mirror Earth: The Search for Our Planet's Twin
Michael D. Lemonick
(Walker Books, 2012)

Oxford Dictionary of Astronomy
ed., Ian Ridpath
(Oxford University Press, 2012)

Planetary Sciences
Imke de Pater & Jack Lissauer
(Cambridge University Press, 2001)

Secrets of the Universe
Paul Murdin
(University of Chicago Press, 2009)

Space: From Earth to the Edge of the Universe
Carole Stott, Robert Dinwiddie & Giles Sparrow
(Dorling Kindersley, 2010)

Strange New Worlds: The Search for Alien Planets and Life Beyond Our Solar System
Ray Jayawardhana
(Princeton University Press, 2011)

Universe
Roger A. Freedman & William J. Kaufmann
(W.H. Freeman, 2010)

Universe: The Definitive Visual Guide
Carole Stott & Martin Rees
(Dorling Kindersley, 2012)

Unveiling the Edge of Time: Black Holes, White Holes, Wormholes
John Gribbin
(Crown Publications, 1994)

WEBSITES

http://www.nasa.gov
The National Aeronautics and Space Administration (NASA) is the US government's agency responsible for the nation's civilian space programme, aeronautics and aerospace research. It shares data with other national and international organizations through its Earth Observing System, Great Observatories and Greenhouse Gases Observing Satellite.

http://apod.nasa.gov/apod
Astronomy Picture of the Day features a different astronomy and space-science related image each day, with a short explanation by a professional astronomer.

http://kepler.nasa.gov
NASA's Kepler Mission website. The space observatory's mission is to search for habitable planets.

http://www.esa.int
The European Space Agency's website. ESA is an intergovernmental organization, currently with 19 member states. It participates in the International Space Station programme, maintains a major spaceport in French Guiana, and is involved in the design of launch vehicles.

http://www.russianspaceweb.com
Russia's Federal Space Agency, commonly called Roscosmos, is the government agency responsible for Russia's space science programme and aerospace research. The above website is in English.

NOTES ON CONTRIBUTORS

EDITOR

François Fressin is a research associate at Harvard University, Massachusetts. Born in Lille, France, where he received a masters degree in Engineering, he then earned a masters and PhD in Astrophysics from the University of Paris. His research focuses on detecting and characterizing planets orbiting other stars. He is a member of the Kepler mission that aims to detect Earth-like planets, possibly suitable for life. Dr Fressin is leading statistical studies to establish the frequency of these distant worlds and the connection to their host stars. He is a founding member of the A STEP project, studying the potential of Dome C, Antarctica, as an astronomical observation site. Using NASA's Kepler space telescope, he was involved in the discovery of the majority of the smallest exoplanets known to date. In December 2011, he discovered the first two Earth-size planets orbiting a star other than the Sun.

FOREWORD

Martin Rees is a Fellow of Trinity College and Emeritus Professor of Cosmology and Astrophysics at the University of Cambridge. He holds the honorary title of Astronomer Royal, served for ten years as director of the Institute of Astronomy at Cambridge and as Master of Trinity College (2004–12). In 2005 he was appointed to the House of Lords, and he was President of the Royal Society for the period 2005–10. He is a foreign associate of the National Academy of Sciences, the American Academy of Arts and Sciences, the American Philosophical Society and an honorary member of the Russian Academy of Sciences, the Pontifical Academy and several other foreign academies. He has been president of the British Association for the Advancement of Science (1994–95) and the Royal Astronomical Society (1992–94). Professor Rees is currently on the Board of the Princeton Institute for Advanced Study, the Cambridge Gates Trust, and has served on many bodies connected with education, space research, arms control and international collaboration in science.

Darren Baskill is an astrophysicist based at the University of Sussex in Brighton, where he manages the outreach programme in physics and astronomy. Dr Baskill is also a freelance astronomer for the Royal Observatory, Greenwich, London, delivering planetarium shows and courses in astrophotography.

Zachory K. Berta studies exoplanets – planets that orbit stars other than our Sun. He is actively engaged in searching for new exoplanets and in observing the atmospheres of these distant worlds, working towards addressing the age-old question: 'Is there life elsewhere in the Galaxy?' Zachory Berta is a graduate student in Astronomy at the Harvard-Smithsonian Center for Astrophysics in Cambridge, Massachusetts.

Carolin Crawford is the Gresham Professor of Astronomy and a Fellow and lecturer at Emmanuel College, Cambridge. During her research career she studied the most massive galaxies in the Universe, located at the cores of clusters of galaxies. She runs the public outreach programme at the Institute of Astronomy in Cambridge, and has given hundreds of public presentations to a wide variety of audiences. In 2009, she received a UKRC award for communication of science. Professor Crawford is a regular contributor to both national and local radio in the UK.

Andy Fabian is the Royal Society Research Professor at the Institute of Astronomy at the University of Cambridge. He leads the X-ray astronomy group, working on clusters of galaxies, black holes and their interrelationship. He was President of the Royal Astronomical Society in 2008–10, and is a Fellow of the Royal Society. His career has ranged from observing the X-ray sky from a rocket launched in the Australian outback at Woomera to mapping the Perseus cluster of galaxies over several weeks, using the Chandra X-ray Observatory. He is looking forward to working on data at the next Japanese-US X-ray Observatory, Astro-H.

Paul Murdin is an astronomer studying supernovae, black holes and neutron stars, working at the Institute of Astronomy at the University of Cambridge, England. In a former life he held influential roles at the British National Space Centre and in the government funding agencies for astronomy in the UK. Dr Murdin pursues a secondary career as a broadcaster, commentator, lecturer and writer on astronomy. He was honoured as an Officer of the Order of the British Empire by the Queen for his work in international astronomy and scientific outreach.

INDEX

ACKNOWLEDGEMENTS

PICTURE CREDITS
The publisher would like to thank the following organizations for their kind permission to reproduce images in this book. Every effort has been made to acknowledge pictures; however, we apologize if there are any unintentional omissions.

Images throughout supplied by ESA/European Space Agency and NASA/courtesy of nasaimages.org

Corbis/Bettmann: 98; Colin McPherson: 66.
Fotolia: 26.
Getty Images/Evelyn Hofer/Time Life Pictures: 140.
Science Photo Library: 30, 81.